M000271968

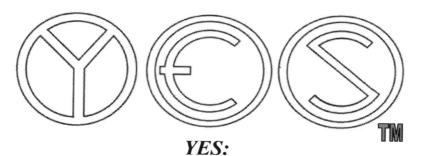

YES:
Young Earth Science and The Dawn of a New WorldView

We appreciate your feedback:
YoungEarthScience@yahoo.com

Please check out the YoungEarthScienceBlog:
http://youngearthscienceblog.blogspot.com/

Follow us on Twitter:
https://twitter.com/youngearthsci

YES: Young Earth Science and The Dawn of a New WorldView (Old Earth Fallacies and the collapse of Darwinism) **by Jay Hall**

ISBN 978-0-692-32007-5

Published by IDEAS (Intelligent Defense of Evolution Alternatives through Science)

IDEAS
POB 732
Big Spring TX 79721

Cover design by Francis Schaeffer Hall

Photo courtesy of Nancy Kitts

Dedication

David B. Kitts (d. 2010), Metageologist, served in New Guinea and the Philippines during World War II. In 1949, he studied Zoology at Columbia University and worked with Theodosius Dobzhansky.[1] He switched to Paleontology and was supervised by George Gaylord Simpson (d. 1984), the most influential paleontologist of the last century. He completed his doctorate in 1953. He was the curator of the University of Oklahoma's Stovall Museum (now the Sam Noble Oklahoma Museum of Natural History). Kitts was a Professor in both the Geology and the History of Science departments. He was the department chair of the History of Science department from 1973 to 1979.[2] He was even an adjunct professor in OU's Philosophy Department! Kitts provided a Philosophy of Geology in *The Structure of Geology* (1977). Despite being a victim of polio, Kitts was an avid cyclist, riding in England, France and Oklahoma.

Dr. Kitts was my History of Science professor. He was a superb instructor and helped hone my critical thinking skills. In 1981, I interviewed him on the topic of evolution – part of this is included in Chapter 3. I greatly admire his bold honesty to discuss challenges to traditional Darwinism and Old School gradualist geological paradigms.

Notes:
1) "David B. Kitts, Metageologist"
http://cas.ou.edu/Websites/oucas/images/hsci/Eloge-DBK(2011).pdf
2) "David B. Kitts (1923-2010)"
http://www.hssonline.org/publications/Newsletter2011/April-in-memoriam.html

Abbreviations and Acronyms

AAPG = American Association of Petroleum of Geologists
a.f. = adapted from
B = billions (of years)
CAT = catastrophe
ETL = Essential Types of Life
EQ = earthquake
FRS = Fellow of the Royal Society
GOT = Global Ocean Theory
IC = Intelligent Catastrophism
K = thousands (of years)
KT = Cretaceous/Tertiary boundary
M = millions (of years)
OEF = Old Earth Fallacy
RPM = Rapid Plate Movement
SERGA = Singular Epoch of Rapid Geologic Activity
YES = Young Earth Science

Contents and Chapters

Introduction

Young Earth Science (YES) is the idea that this planet is thousands of years old rather than billions of years in age. Who in the world would hold to such a view? Are YES advocates wacky ideological nuts? Science is based on observational evidence and experiment, yet the mainstream scientific consensus often changes over time (e.g. plate tectonics). Truly accurate science favors YES. Did you create the cosmos? Then, how do you know that the Earth is billions of years old? Do you abdicate your responsibility to think critically by trusting the experts? We have not yet invented a History Observation Device (HOD) to prove that the Earth is really billions of years old. Dr. Who has left the building along with the Universal Historian.

Who in this list supports YES?
  Will Smith
  Steven Spielberg
  Barack Obama
  Kirk Cameron
  Oprah
  Kim Kardashian
  Al Gore
The man in the middle is YES friendly, of course, but what about Al Gore? An op-ed on Al Jazeera opposed evolution, which is a key aspect of Earth's age, and Al Gore sold his Current network to Al Jazeera.[1] Origins research makes for strange bedfellows.

Time.com

Old Earth Fallacies (OEF's) and evolution are closely connected. Is there a well-known scholar who opposed evolution? Philosopher Mortimer Adler (d. 2001) was the Associate Editor of *The Great Books* series and co-author of the popular *How to Read a Book*. Adler was on the cover of *TIME* on March 17, 1952.[2] Adler was also the chairman of the Board of Editors of the *Encyclopedia Britannica* for some time. You can read about Adler's opposition to evolution in *The Difference of Man and the Difference It Makes*. I will show in chapter five that anti-evolution views leads to YES.

Jeremy Rifkin (*Time Wars*), president of the Foundation on Economic Trends, has advised numerous European heads of state. Rifkin wrote *Algeny* which is described on the cover as a "provocative critique of Darwinism." One section is titled,

"The Darwinian Sunset: The Passing of a Paradigm." Norman Macbeth (*Darwin Retried*) notes that, "Rifkin asserts at length that Darwinism is a moth-eaten theory that should be discarded once and for all."[3]

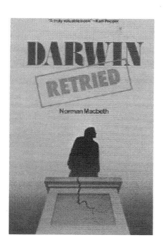

Is Young Earth Theory merely a fringe movement? - not at all. In the land of Darwin, Darwinism is going down (as in Down House). A mere 37% of the British think that evolution is beyond reasonable doubt. A full 32% favor the Young Earth view which sees this planet as thousands of years old (not billions). This 2009 UK poll involved 2,000 people.[4]

a.f. telegraph.co.uk

WorldView Matters

Determining the age of the world is based on many unproven PreSuppositions. It is like Sherlock Holmes trying to solve a crime. There are clues, but Holmes was not there to see the murder. Even with multiple witnesses there may be apparent contradictions. In truth, there is a crisis of confidence in the persuasive power of OEF's. I have a thousand ideas why going for the Billions is a dead end. Think young – think YES! Will npr provide a fair and balanced treatment of YES? Science Friday, your day has come to discuss OEF's.

a.f. sfmconsulting.org

Milford Wolpoff is a paleoanthropologist (Museum of Anthropology, University of Michigan) and said,

> I believe a philosophical framework is not something that can be eliminated in order to provide "objectivity." In my view, "objectivity" does not exist in science. Even in the act of gathering data, decisions about what data to record and what to ignore reflect the philosophical framework of the scientist.[5]

So, one's WorldView has a great impact on how we interpret the data we observe.

Geophysicist Pascal Richet pointedly observes in *A Natural History of Time* that, "Obviously it is not easy to consider the question of time from beyond the context of all philosophical or metaphysical presuppositions."[6] Geochronology, origins and worldview are intimately connected. In 1981, John Greene wrote *Science, Ideology, and World View: Essays in the*

History of Evolutionary Ideas. One chapter is titled "Darwinism as a world view."

Your WorldView represents your most basic assumptions regarding how you understand the world, history and life in general. My father, Francis "Dick" Doyle Hall went through the Depression, the dustbowl and went to California to find work – he was a genuine Okie. The stories he told me make history more real. Albert Henry Woolson (d. 1956) was the last surviving Civil War vet and died at the age of 109. Theoretically, my father could have met Woolson and heard his stories of the war.[7] Many cultures place an emphasis on time. The Mayans may have been wrong about the end of the world, but they calculated the solar year to astonishing accuracy – they were off by only 23 seconds![8] History is precious and short (K's not B's). Should we be wasting time by binging on reality TV?

Commenting on the BBC radio program "Ageing the Earth," Barbara Cordner had this insight:

> The concept that humans have only been on earth for such a short time [v. Earth's 4.5B years], and that should we become extinct there would be very little record in comparison with other lifeforms served to put our little squabbles and wars in real perspective.[9]

If history is important and the Earth is only thousands of years old (not billions), then our ancestors' significance changes and they were more than just a speck among the billions of years of prehistory. Consider the words of Walter Bedeker from the

Twilight Zone ("Escape Clause" episode), "The world goes on for millions of years and how long is a man's life? … a drop, a microscopic fragment."[10] WorldView matters!

Francis Haber, in his book on the age of the world, published in 1959 (the centennial of Darwin's *Origin of Species*), said,

> … the great revolution in concepts of time was undoubtedly in the twentieth century. Sweeping through early twentieth century thought like a tidal wave, <u>time saturated physics, cosmology, philosophy, psychology, literature, and art, to such an extent that the world outlook [WorldView] of Western man was thoroughly transformed</u> to include a fourth dimension.[11]

Geneticist Richard Goldschmidt (d. 1958) provides a wise statement that relates to the importance of WorldView and PreSuppositions:

> The development of the evolutionary theory [or OEF's] is a graphic illustration of the importance of the *Zeitgeist* [general beliefs of a certain time]. A particular constellation of available facts and prevailing concepts dominates the thinking of a given period to such an extent that it is very difficult for a heterodox viewpoint to get a fair hearing. …The fact that the synthetic theory [of evolution] is now so universally accepted is not in itself proof of its correctness.[12]

The fact that OEF's are now so universally accepted is not in itself proof of their correctness! Goldschmidt was right! The Geo intelligentsia might be wrong about this planet's birthday. Stephen Jay Gould plainly stated that Charles Lyell (d. 1875) had a bias against catastrophism: "*<u>Principles of Geology</u>* [Lyell's magnum opus] <u>is a brief for a world view</u> – time's stately cycle as the incarnation of rationality."[13]

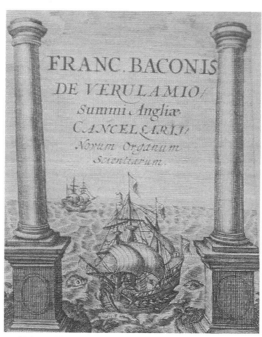

wikipedia

We are often bound by our PreSuppositions. Pioneer science advocate Francis Bacon (d. 1626) warned of "idols which have crept into men's minds from the various dogmas of peculiar systems of philosophy" and that the mind "supposes a greater degree of order and equality in things than it really finds" and thus we must avoid "preconceived fancies."[14]

"Catastrophism" is a Real Word

We show in chapters 4 and 5 that there is a close connection between catastrophism (most of the rock record formed quickly) and YES. Has the mainstream ever supported catastrophism? Science journalist Fred Warshofsky was not part of the scientific establishment, but he won two Emmys and a Lasker Science Award. Warshofsky wrote an entire book defending catastrophism, *Doomsday – The Science of Catastrophe* which was even made into a movie hosted by Vincent Price.[15] Warshofsky was the science editor of "The 21st Century" (CBS) and the director and writer of "In Search of Ancient Mysteries" (NBC) hosted by Rod Serling.[16] Antony

Milne has written a sort of sequel, *Doomsday: The Science of Catastrophic Events*.

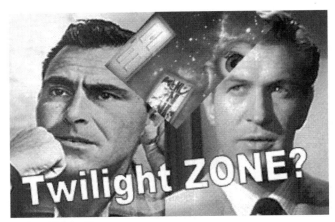

a.f. wikipedia

According to the *Oxford English Dictionary*, "Catastrophism" means, "The theory that certain geological and biological phenomena were caused by catastrophes or sudden and violent disturbances of nature, rather than by continuous and uniform processes."[17] Catastrophism has been scorned and denigrated in the past, but it is now experiencing a resurgence.

YES we can!

Scientists often reject YES before looking at the supporting arguments. As Paul Forman observes, "… I take as the defining characteristic of the <u>immediate</u> reception of innovations: people, and physicists, not only *tend to find what they are looking for, but also fail to recognize what they are not prepared to see*."[18]

Where do we place the beginning of modern science? Some might start in 1543, when Copernicus proposed the heliocentric theory. Others might place it at the publication of *Novum Organum Scientiarum* by Francis Bacon (1620). We will use the latter figure. Let's take the conclusion of Haber above and date the dominance of the Old Earth view at 1900. For hundreds of years most scientists held that the earth was young and not billions of years old. That is, for about 71% of the history of science a considerable number of scientists supported YES. Could they all have been wrong? Consider the evidence presented in this book and escape from the Deep Time Cult. The Scientific Research Complex is in the tank for OEF. You can be an agent of change and break the status quo.

According to Boyce Rensberger,

> Until Lyell published his book [*Principles of Geology*], most thinking people accepted the idea that the earth was young, and that even its most spectacular features such a mountains and valleys, islands and continents were the products of sudden, cataclysmic events...[19]

John Valley, an editor with *Elements*, observes that, "The 'Age of the Earth' is one of the most common titles in the geological literature, and with good reason. The scientific and philosophical implications are immense."[20]

Francis Schaeffer Hall

Carl Sagan (above) in *Contact* proposed communication with alien intelligence. What if occupants of a UFO said the earth was thousands of years old? Would the OEF hawkers believe them? What if records from an advanced civilization from another planet found in an archeological dig supported Young Earth Theory. How would Big Geology respond?

Charles Darwin (d. 1882), himself, provides us with some helpful advice on the age of the earth controversy:

A fair result can be obtained only by fully stating and balancing the facts on both sides of each question... - Charles Darwin

Notes:
1) "Are humans created or evolved?" by Muhammad Abdul Bari, http://www.aljazeera.com/indepth/opinion/2013/01/2013122111010987332.html
2) Adler on TIME, Mar. 17, 1952, http://topics.time.com/mortimer-adler/covers/
3) "*Algeny* - Additional Reviews" http://www.foet.org/books/algeny-reviews.html
4) "Rescuing Darwin" by Nick Spencer and Denis Alexander, http://campaigndirector.moodia.com/Client/Theos/Files/RescuingDarwin.pdf
5) Wolpoff, M. 1999. *Paleoanthropology*. 2nd ed., Boston, MA: McGraw-Hill, p. iv, emphasis added.
6) *A Natural History of Time* by Pascal Richet, translated by John Venerella (University Of Chicago Press, 2010), p. x.
7) "Albert Woolson"

http://en.wikipedia.org/wiki/Albert_Woolson

8) "Mayan calculations hold up, to a point" by Tony Rice
http://www.wral.com/mayan-calculations-hold-up-to-a-point/11908656/

9) "In Our Time – Debate"
http://www.bbc.co.uk/radio4/history/inourtime/inourtime_comments_agein
g_earth.shtml

10) "The Twilight Zone S01 E06 Escape Clause"
http://www.youtube.com/watch?v=neIUJg-d3Vs

11) *The Age Of The World* by Francis Haber (Johns Hopkins Press,
Baltimore, MD, 1959), p. 8, emphasis added.

12) Quoted in *The Science of Evolution* by William Stansfield (Macmillan,
NYC, 1977), p. 577.

13) Quoted in *Time's Arrow, Time's Cycle* by Stephen Jay Gould (Harvard
Univ. Press, Cambridge, MA, 1987), p. 143, emphasis added.

14) *Great Books of the Western World* ed. by Robert Hutchins, Assoc. Ed.
Mortimer Adler (Encyclopedia Britannica, Chicago, 1952), Vol. 30, pp.
110, 111, *Novum Organum* by Francis Bacon (1620), First Book,
Aphorisms 44, 45 and 54.

15) "Days of Fury (Narrated by Vincent Price)"
http://www.youtube.com/watch?v=WwNwm1lctcU

16) "In Search of Ancient Mysteries"
http://www.youtube.com/watch?v=ictEQa__4sY

17) "Catastrophism"
http://www.oed.com/view/Entry/28797?redirectedFrom=catastrophism
Accessed 7/5/2013.

18) "The Reception of an Acausal Quantum Mechanics in Germany and
Britain" by Paul Forman in The Reception of Unconventional Science, ed.
by Seymour Mauskopf (Westview Press, Boulder, CO, 1979), AAAS Select
Symposium, No. 25, pp. 11-50, p. 11, italics added.

19) Quoted in *Time's Arrow, Time's Cycle* by Stephen Jay Gould (Harvard
Univ. Press, Cambridge, MA, 1987), p. 114, emphasis added.

20) "The Age of the Earth" by John Valley, *Elements*, Feb. 2013, p. 3,
emphasis added.

Chapter 1
The Real Age of the World

Cleinias: What do you mean [about the origin of government]?
Athenian Stranger: I mean that he might watch them from the point of view of time, and observe the changes which take place in them during infinite ages.
Cleinias: How so?
Athenian Stranger: Why, do you think that you can reckon the time which has elapsed since cities first existed and men were citizens of them?
Cleinias: Hardly.
Athenian Stranger: But you are sure that it must be vast and incalculable?
Cleinias: Certainly.[1]
 - Plato, *Laws,* Book 3

Man has been here 32,000 years. That it took a hundred million years to prepare the world for him is proof that that is what it was done for. I suppose it is. I dunno. If the Eiffel Tower were now representing the world's age, the skin of paint on the pinnacle knob at its summit would represent man's share of that age; and anybody would perceive that the skin was what the tower was built for. I reckon they would, l dunno.[2]
 - Mark Twain

Conventional science says this planet is 4.5B years old. Former aeronautical engineer and award winning SciFi author, James Hogan states that conventional dating was, "… manufactured to provide the long time scales that Lyell and Darwin needed."[3]

Are we really certain that the Earth is billions of years old? Lord Ormathwaite, in his *Astronomy and Geology Compared* (1872) argued,

> Mr. Darwin requires for the development of his theory <u>enormous periods of time, far exceeding any of which we have the slightest knowledge;</u> this alone <u>places his whole system beyond the domain of fact and in the regions of mere reverie and imagination.</u>[4]

Darwin's assumption of hundreds of millions of years rested on a weak foundation.

Scottish geologist Sir Archibald Geikie admitted that,

> Until Darwin took up the question, the <u>necessity for vast periods of time,</u> in order to explain the characters of the geological record, was <u>very inadequately realized</u>. … no one before his day had perceived how enormous must have been the periods required for the deposition of even some <u>thin continuous groups of strata</u>. He supplied a criterion by which, to some degree, the relative duration of formations might perhaps be apportioned. When he declared that the intervals that elapsed between <u>consecutive formations</u> may sometimes have been of far longer duration than the formations themselves, <u>contemporary geologists could only smile incredulously in their bewilderment</u> …[5]

Evolution requires billions of years, but what if we <u>don't</u> have an Old Earth. The rocks said "little time," but Darwin said "much more time."

Why should we think the earth is millions or billions of years old? According to the *New International Encyclopedia* (1915 ed.),

> [An] evidence pointing to a great age for the earth is supplied from a study of the fossils preserved in the rocks. The evolution of plant and animal life seems, in general, to have been gradual … all lines of evidence

agree in pointing to the conclusion that geological time is to be reckoned in millions of years ...[6]

Evolution proves millions of years?! Isn't that putting the cart before the horse? All evidence points to an old earth? This was written in 1915 when gradualistic uniformitarianism was dominant. Now that we are in the Renaissance of Catastrophism, should we not jettison the millions of years?

Philosopher David Hume supported Young Earth Science (YES):

> It is not two thousand years since vines were transplanted into France, though there is no climate in the world more favorable to them. It is not three centuries since horses, cows, sheep, swine, dogs, corn, were known in America. Is it possible that during the revolutions of a whole eternity, there never arose a Columbus, who might open the communication between Europe and that continent? We may as well imagine, that all men would wear stockings for ten thousand years, and never have the sense to think of garters to tie them. All these seem convincing proofs of the youth, or rather infancy, of the world; as being founded on the operation of principles more constant and steady, than those by which human society is governed and directed. Nothing less than a total convulsion of the elements will ever destroy all the European animals and vegetables, which are now to be found in the Western world.[7]

Hume saw that the rapid rise of technology and the progress of global travel favor the young earth view. However, if there is an Earth-reboot via a global natural disaster, then much of the Earth would appear new and young.

James Hogan said, "A comparatively young world – in the sense of the surface we observe today – is compatible with unguided Catastrophist theories..."[8]

Lucretius (d. 45 BC), Roman poet and Epicurean, advocated YES (*On the Nature of the Universe*, Book 5, 326):

> Why have no poets sung of feats before the Theban War and the tragedy of Troy [~1200 BC]? The answer, I believe, is that the world is newly made: its origin is a recent event, not one of remote antiquity. That is why even now some arts are being perfected.[9]

If human history truly goes back tens of thousands of years, where are the epic poems honoring the great feats of those heroes who lived during the distant past? Technology is still improving. Could this be an indicator that civilization is not that old?

Yes Virginia, There is Historical Science

Geology as a historical science is different than chemistry and physics. "Dr. Wow" can put his/her hand into a container of molten lead, but a geologist can't re-create the breakup of Pangea in the lab. So, geoscientists should be humble when they attempt to measure the age of the earth. Some deny the significance of historical science, yet paleontologist George Gaylord Simpson wrote a whole chapter on "Historical Science" in the book *The Fabric of Geology*!

In their paper on "Geology as an Historical Science," Jeff Dodick and Nir Orion affirm,

> There are many reasons why the temporal revolution of the earth sciences has been, for the most part, neglected. Geology is often thought to have many practical and theoretical limitations which 'undercut its claim to knowledge' [Robert Frodeman] …Such problems include incompleteness of the stratigraphic and fossil records; lack of experimental verification; and the nature of 'deep time' itself, which precludes direct observation …[10]

How Old is *Your* Planet?

Around 1900 there was a wide range of estimates for the earth's age that were seriously considered. Hermann von Helmholtz gave a date of 22M years based on the sun's loss of energy. John Joly gave a date of about 90M years based on the amount of salt in the ocean. By 1931 Arthur Holmes used the results of radiometric dating to assert that this planet is at least 1.5B years in age.[11] From 22M to 1.5B, that's a 68-fold increase! Here is the latest version of the geologic Periods:

Geologic Time Scale	
Start of	M's Years
Quaternary	2
Tertiary	66
Cretaceous (K)	146
Jurassic	200
Triassic	251
Permian	299
Carboniferous	359
Devonian	416
Silurian	444
Ordovician	488
Cambrian	542

a.f. usgs.gov

Ancient Fossils Look Young

Hadrosaur skin was found in Alberta, Canada in 2012. The duck-billed hadrosaur is supposedly 65M years old. University of Regina physicist Mauricio Barbi excavated the fossil.[12] How can real skin last 65M years?!! If the earth is only thousands of years old, then dino skin is not a problem.

In 2012, Mary Schweitzer, paleobiologist at North Carolina State University, found proteins in a dinosaur fossil! DNA is thought to have a half-life of 521 years.[13] If YES is true this is not an issue, but if these dinos are really over 60M years old, how could the DNA survive that long?

a.f. researchnews.osu.edu

In 2006, Ohio State University geologists William Ausich and Christina O'Malley isolated the oldest complex organic compounds found in 350M year old crinoids (sea lilies).[14] Could this be evidence that the crinoids are really not that old, since organic molecules tend to break down?

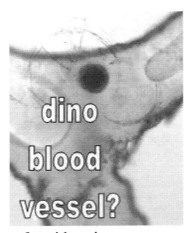

a.f. smithsonianmag.com

Proteins have been found in a 150M year old Seismosaurus from the Morrison formation (New Mexico)!![15] Could these dino proteins really be 150M years old? Proteins naturally break down over time. Proteins in dinos and Old Earth Fallacies (OEF's) don't mix. Mary Schweitzer found blood vessels in 68M *T-Rex* fossils! Could such dino soft tissue really survive for 68 million years old?[16] Writing in *Nature* online, Ed Yong states that, "Theoretical predictions and lab experiments suggest that proteins cannot survive for more than a few million years."[17] Does that mean that Schweitzer's dino soft tissue is only 3M or 6M years old and NOT 68M years old?

In 1992, Rob DeSalle et al announced this amazing discovery:

> DNA was extracted from the fossil termite *Mastotermes electrodominicus* preserved in Oligo-Miocene amber (25 million to 30 million years old). Fragments of mitochondrial ... and nuclear ... genes were amplified by polymerase chain reaction. ...The fossil termite shares several sequence attributes with *Mastotermes darwiniensis*. [This discovery is] the oldest DNA yet characterized.[18]

The soft tissue of brachiopods have been found in the Silurian Herefordshire (425M) in the UK. A pedicle with distal rootlets along with a lophophore are preserved.[19] Is it not more reasonable to conclude that the dating methods are in error, than to think soft tissue can survive for 425M years?! At 10°C (average UK temperature) collagen will decompose to just 1% of the original concentration between 200K and 700K years!![20] Could these fossils with proteins really be millions of years old if collagen disintegrates so fast?

Allegedly 100M year old trees have been found in northern Alberta. They are not petrified and the cell walls are clearly seen. Could this nearly perfect preservation really last for millions of years?[21]

Racemization Dating

Life overwhelmingly has exclusively left-handed amino acids. Over time, these turn into a 50-50 mixture of left-handed and right-handed amino acids. This process is called racemization. Geoffrey Eglinton and Mary T. J. Murphy admit,

> …there have been a number of studies of <u>amino acids</u> in generally random samples of geologic interest. Samples representing <u>most of the geologic column including the Precambrian</u> have been reported to contain several <u>amino acids including some relatively unstable ones</u>.[22]

Eglinton and Murphy claim that these unstable non-racemized amino acids are due to "Problems of Contamination."[23]

John Wehmiller and P. E. Hare report that,

> Amino acids in sediments show an initial <u>rate of racemization almost an order of magnitude faster than the rate observed for free amino acids</u> at a comparable pH and temperature. …<u>Isoleucine is racemic in samples older than about 15 x 10^6 years</u>.[24]

If amino acids racemize in just 15M years and non-racemized amino acids are found throughout the geologic column, then the established geologic timescale of hundreds of millions of years is wrong! Calling it "contamination" is just an excuse.

Tree Rings and YES Ring True

What tree in the fossil record has the most rings? There is at least one permineralized Sequoia specimen with 815 growth rings.[25] The logic of YES is easy. It's just common sense – if the rocks are millions of years old, where are the fossil trees that are thousands of years old? A number of trees are quite old: Sequoias, Redwoods, Bristlecone pines, Junipers and Olive trees. Why are there no 2K, 3K or 4K year old trees in the fossil record? Maybe YES <u>is</u> true?

a.f. nrmsc.usgs.gov   a.f. usgs.gov

The oldest forest trees date to 385M years.[26] If the rock record of trees represents hundreds of millions of years, then we should find trees with 3,000 rings or 5,000 rings. Where are they?! Vice-Admiral Robert FitzRoy (d. 1865) was captain of HMS Beagle during Charles Darwin's eventful voyage. FitzRoy has a tree with great longevity named after him from the cypress family, *Fitzroya cupressoides*, which can live over 3600 years!![27]

Tree	Age
Pinus longaeva	4844 years
Fitzroya cupressoides	3622 years
Sequoiadendron giganteum	3266 years
Juniperus occidentalis var. australis	2675 years
Lagarostrobos franklinii	ca. 2500 years

This chart documents some of the oldest trees.[28]

Lagarostrobos franklinii is known as the Huon pine or Macquarie pine, but it's really a podocarp rather than a pine.[29]

The Population Bomb favors YES

It is generally agreed that Neanderthals lived 300K years ago. Mainstream anthropologists say that they became extinct around 25K BC.[30] In the early 1960's the world population growth rate was 2.2% per year. In 2009, the growth rate was 1.1%.[31]

The basic population equation is $P(t) = P(0)e^{rt}$ where r is the growth rate. So let's take just <u>one percent</u> of the 2009 growth rate and assume that Neanderthals roamed the earth for 275K years with an initial population of two. We are using such a miniscule growth rate to account for diseases and natural disasters. The gives the following population size:

$$P(275K) = 2[e^{\wedge}(.00011*275K)] = \textbf{27 Trillion!!}$$

a.f. wikipedia

Where are all the Neanderthal cemeteries? According to Antonio Rosas (Museum of Natural History, Madrid), Neanderthals, "… looked after the sick, buried their dead, … decorated their bodies … [and practiced] self-medication." At a burial site in Sima de las Palomas (Spain), three Neanderthals have their hands raised in the same manner. Archaeologist Michael Walker thinks this indicates ritual burial and suggests that Neanderthals believed in an afterlife.[32]

Neanderthals practiced ritual burial – so where are the trillions of graves? They had a very wide geographic distribution, so we should find Neanderthal bones by the truck load.[33] Or, could the dating methods be wrong and the Earth young?

Who's Your Common Ancestor?

Based on math models of human migration across the globe Douglas Rohde et al conclude, "… the genealogies of all living humans overlap in remarkable ways in the recent past. In particular, the MRCA [Most Recent Common Ancestor] of all present-day humans lived just a few thousand years ago in these models."[34] Is humanity only thousands of years old? Is the Earth itself only thousands of years old?

Most mutations are neutral or harmful. Mutations have a cumulative effect and so have a hurtful result on the human population as a whole. Michael Nachman and Susan Crowell writing in the journal *Genetics* said,

> …we estimate that the genomic deleterious mutation rate (U) is at least 3. This high rate is difficult to reconcile with multiplicative fitness effects of individual mutations… The high deleterious mutation rate in humans presents a paradox. If mutations interact multiplicatively, the genetic load associated with such a high U would be intolerable in species with a low rate of reproduction… [given this scenario] each female would need to produce 40 offspring for 2 to survive and maintain the population at constant size. This assumes that all mortality is due to selection and so the actual number of offspring required to maintain a constant population size is probably higher.[35]

Since humanity's genetic load is so heavy (our genetic purity – less defects and disease - gets worse over time), how can mankind be millions of years old?

In 1991, nine skeletons in Siberia were excavated which included the last Russian Tsar, Nicholas II, his family and servants who were shot by a firing squad in 1918. This research implied a faster mutation rate for mitochondrial DNA (mtDNA) than previous estimates. Writing in *Science*, Ann Gibbons reveals a surprising result:

> Regardless of the cause, evolutionists are most concerned about the effect of a faster mutation rate. For example, researchers have calculated that …the <u>woman whose mtDNA was ancestral to that in all living people</u> - lived 100,000 to 200,000 years ago in Africa. <u>Using the new clock, she would be a mere 6000 years old.</u>[36]

Where in the World is Thor Heyerdahl?

The global distribution of mankind did not take hundreds of thousands of years. Thor Heyerdahl (d. 2002), the famous Norwegian explorer, travelled 5,000 miles in his raft, Kon-Tiki, from Peru to the Tuamotu Islands in 1947. He thus proved that Polynesian-American contact was possible.[37] There is genetic evidence that supports the Peru-Polynesian connection. A ship named Ra II built from totora sailed from Morocco to Barbados thus demonstrating the possibility of Africans reaching the Americas. The film *Ra* documented Heyerdahl's expeditions.[38]

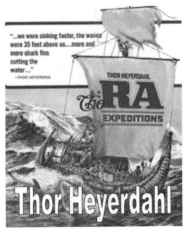

a.f. wikipedia

In 1998, *The Oregonian* carried this provocative headline: "America's early people may have come by boat from Asia." As Richard L. Hill reported,

> New evidence suggests the earliest people in the Western Hemisphere may have arrived by boats along coastal routes from Asia to North America rather than just across an ancient ice bridge that linked the two continents. They also may have arrived earlier than previously thought.... C. Loring Brace, [is] a professor of anthropology at the University of Michigan. An expert on skull and facial features, he said some of the remains he's studied appear to be related to the ancient Ainu of Japan, and others are related to people in northeast Asia, southeast Asia and China.[39]

Archaeologist Anton Wilhelm Brogger aroused quite a controversy when at an academic conference in Oslo in 1936 he advocated his theory of a golden age of navigation. This period where the entire globe was explored reached its acme around 3000 BC and was on the decline by 1500 BC. If Brogger was correct, it did not take many thousands of years for man to span the globe.[40]

Charles Boland wrote *They All Discovered America* in 1961. Boland investigated clues that show that the Phoenicians, Romans, Chinese, Irish and Vikings came to America before Columbus.

Maps of the Ancient Sea Kings

The Oronteus Finaeus map of 1532 shows Antarctica with mountains! The interior is free of rivers and mountains and so may have been covered with ice. Could this map be based on information from before the Ice Age? If so, the Ice Age was not that long ago.[41]

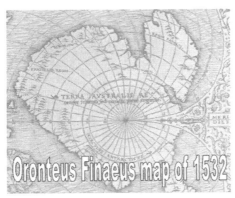

a.f. *Maps of the Ancient Sea Kings*

Philippe Buache (d. 1773) made a map of Antarctica without ice! It was originally published in 1739. This map was apparently based on a work that was created before the Ice Age began.[42] Compare Buache's map with the sub-glacial topographical map of Antarctica. Was it possible to navigate through the middle of the southernmost continent at one time?

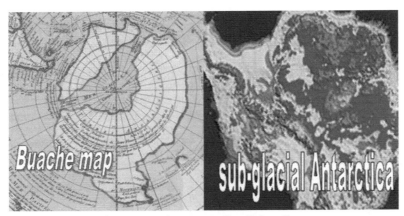

a.f. libweb5.princeton.edu   a.f. wikipedia

In the Zeno Map of the North published in 1558 and based on a 1380 map left by two ancestors, Greenland has no ice! The map has mountains which is correct.[43] Also, there is indeed a flat area within Greenland which is also the case. Could this work be founded on a map drawn before the Ice Age?

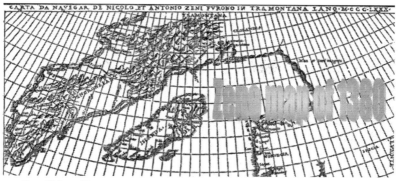

a.f. wikipedia

a.f. *Maps of the Ancient Sea Kings*

The Ibn Ben Zara map of 1487 shows the Aegean Sea with more islands than we see now.[44] Could this be due to the fact that the original document on which this map is based was created during the Ice Age when the sea level dropped? Do these maps indicate that the Ice Age was not that long ago, but actually within human history?

How far did the Chinese extend their global exploration? Gavin Menzies claims China found America in 1421. Dr. Hwa-Wei Lee, Former Chief, Asian Division, Library of Congress (Retired); Dean Emeritus, Ohio University Libraries gives this endorsement of *Secret Maps of the Ancient World* by Charlotte Harris Rees:

> There is much evidence that Chinese were in America hundreds if not thousands of years before Columbus. Based on the rare Asian maps collection of her late father, Dr. Hendon M. Harris, the author has painstakingly researched, including using the resources of the Library of Congress, to present her findings that Chinese had indeed traveled by sea to the Americas since 2000 BC.[45]

Here is one argument Rees makes for early Chinese influence in the Americas:

> Dr. H. Mike Xu … who identified ancient Chinese writings on Olmec celts in Mexico and on other rocks in the United States. He wrote *Origin of the Olmec Civilization* (University of Central Oklahoma Press). The Olmec culture began about 1200 B.C. Xu contends that the Olmec were transplants from China. *The U.S News & World Report,* November 4, 1996 reports that the incisions on Olmec artifacts have been verified by other experts from China as being Shang era Chinese writing.[46]

a.f. uscpfa.org

An old Chinese map on the cover of *US-China Review* is titled *Everything under Heaven.* China and Korea are at the center. Asia arched round to Alaska and down to North America, which was labeled "Fu Sang." The *Shan Hai Jing* (Book of Mountains and Seas) demonstrates that Fu Sang is America.[47]

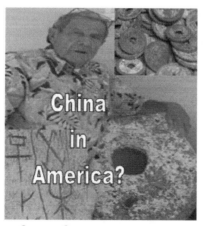

a.f. uscpfa.org

Bob Meistrell, founder of BodyGlove (wet suits), found around 30 Chinese style anchors off the California coast in 2010. They have been dated to 3K- 4K years old. Chinese coins found in a Canadian creek bear a date of about 1200 BC! A tree in North America has Chinese oracle bone script (c. 1300 BC).[48]

History is pro-YES

Native Americans told James Wilkinson (governor of Louisiana Territory in 1805) that a volcano was in the Yellowstone area. The last lava flow was supposedly around 70,000 BC by generally accepted dating. Could it be that the dating methods are wrong and Yellowstone experienced eruption much more recently?[49]

Joe Burchfield, the author of *Lord Kelvin and the Age of the Earth*, stated, "I will argue that the invention of geological time involved at least five essential steps ... [including] the acceptance of a terrestrial age significantly greater than the historical record of humankind (the notion, however vague, of 'deep time') ..."[50] Let's not dismiss the possibility that the age of the earth is on the same order of magnitude as human history. If the geologic timescale was invented, is it really true? James Watt (d. 1819) invented a better steam engine, but is it true? Does an invention speak true truth?

The Burzahom site (Period I) in the Kashmir valley in India is dated by radiocarbon to around 2920 to 2550 BC.[51] Wikipedia places the first dynasty of India (Brihadratha) to around 1700 BC.[52] Deng Yinke (*Ancient Chinese Inventions*) places the beginning of the Xia Dynasty at 2070 BC.[53]

Simplicius reports that astronomical observations encompassing 1903 years before the taking of Babylon by Alexander the Great in 331 BC. Thus, 2234 BC was the founding date for Babylon.[54] According to the "short chronology" Sargon, the first ruler of the Akkadian Empire, reigned from 2270 - 2215 BC. The First Babylonian Dynasty started in 1830 BC by this view.[55]

Regarding Egyptian origins, Isaac Preston Cory concluded that, "... Constantinus Manasses ... asserts that the Egyptian kingdom lasted 1663 years, which, added to the year, B.C. 525, in which Cambyses reduced the kingdom to a province of Persia, places its foundation at B.C. 2188 ..."[56]

English Egyptologist David Rohl (*Pharaohs and Kings*) advocates the change of conventional dates for the Egyptian chronology. He wants to modify the dates up to 350 years, making them more recent. For example, Amenemhatl began his reign in 1800 BC according to Rohl, whereas the traditional date is 1985 BC. Professor Amélie Kuhrt, head of Ancient Near Eastern History at University College London, admits that, "Many scholars feel sympathetic to the critique of weaknesses in the existing chronological framework …"[57]

a.f. wikipedia

The Awan Dynasty was the first dynasty of Elam, thought to be rivals of nearby Sumer.[58] The first king of the Awan Dynasty, Peli, ruled from around 2500 BC.[59] Tudiya, the first of the "Kings who lived in tents" in the Assyrian King list, is estimated to have reigned in the 23rd century BC.[60]

a.f. wikipedia

Gilgamesh was the fifth king of Uruk (Iraq) and ruled around
2500 BC. He went on a trek to find Utnapishtim who survived
a global catastrophe.[61] Joshua Mark details the support for the
historicity of Gilgamesh:

> Historical evidence for Gilgamesh's existence is found
> in inscriptions crediting him with the building of the
> great walls of Uruk (modern day Warka, Iraq),
> references to him by known historical figures of his
> time (26th century BCE) such as King Enmebaragesi of
> Kish and, most recently, by the claim of a German team
> of Archaeologists to have discovered the tomb of
> Gilgamesh in April of 2003.[62]

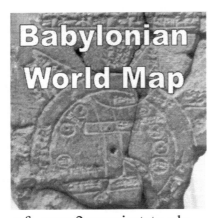

a.f. www2.econ.iastate.edu

The Babylonian map of the world is dated around 700 – 500
BC.[63] The associated text mentions Utnapishtim along with
Sargon of Akkad.[64]

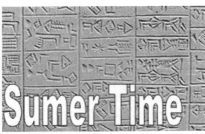

a.f. wikipedia

The earliest historical records of Sumer date to around 2700 BC.[65]

Sumerian King List

King	Reign	Base 60			Decimal Reign
		(60^2)'s	60's	1's	
Alulim	28,800	8	0	0	800
Alalngar	36,000	10	0	0	1,000
En-men-lu-ana	43,200	12	0	0	1,200
En-men-gal-ana	28,800	8	0	0	800
Dumuzid	36,000	10	0	0	1,000
En-sipad-zid-ana	28,800	8	0	0	800
En-men-dur-ana	21,000	5	50	0	550
Ubara-Tutu	18,600	5	10	0	510
				Total	6,660

The Sumerian King List has eight kings before the worldwide cataclysm that the epic of Gilgamesh relates. Sumerian mathematics was based on 60 (sexagesimal), whereas we use the decimal system (base 10). If we convert the reigns from base 60 to decimal, we can see that the Sumerian King List sums to about seven millennia. Notice how the preponderance of even multiples of sixty and 3600 (sixty squared) favors this theory. Could this provide support for the claim that the earth itself is only thousands of years old, and not billions?[66]

Natural Dating Methods say YES!

Richard Milton, Mensan and science journalist, explains the radiocarbon budget,

> After the Earth was formed and acquired an atmosphere, there would be a 30,000 year transition period during which carbon 14 would be building up. At the end of that period, the amount of carbon 14 created by cosmic radiation will be balanced by the amount of carbon 14 decaying away to almost zero. To use Libby's terminology, at the end of 30,000 years, the

terrestrial radiocarbon reservoir will have reached a steady state.[67]

The data indicate that the rate of radiocarbon production rate exceeds the decay rate by as much as 25 percent!! Does that not imply that the Earth is less than 30K years old?[68]

If the world is actually young, how can Australia's Great Barrier Reef, which is supposedly 20K years old, be explained? The coral Acropora cervicornis can grow as fast as 26.4 cm/year!![69] The CRC Reef Research Center dates the living structure of the Great Barrier Reef at less than 8K years.[70] In 1890, just north of the Australian coast, the liner *Quetta* sank because it hit a reef. This part of the ocean had been surveyed between 1802 and 1860 and was considered safe for ships. Could this be an indication of rapid reef growth?[71] Gerald Friedman gives testimony that carbonates can form rapidly:

> Lithification is a rapid process, but how rapid is rapid? A year after one of my visits to Joulter's Cay (Bahamas) I found a sardine can from my previous visit. The cementation of carbonate particles in and around the can was surprising. 382 g of ooids and skeletal material had filled the can and lithified therein, or had become cemented to the outside of the can. The rapidity of marine carbonate cementation implies that … carbonate sediments are subject to marine lithification which retards erosion.[72]

Don't geomagnetic reversals prove that the Earth must be ultra-ancient? A lava flow dated at 15.6M in the Sheep Creek Range (north central Nevada) gives evidence that the Earth's magnetic field can flip rapidly (South-to-North as the compass points). Scott Bogue and Jonathan Glen explain,

> … [this research] reinforces the controversial hypothesis that geomagnetic change during a polarity reversal can be much faster than normal. … the 53° change from east-down to north-down field occurred at

an <u>average rate of approximately 1°/week, several</u> <u>orders of magnitude faster than typical of secular</u> <u>variation</u>.[73]

Geologist Jon Erickson states, "Some geomagnetic reversals, whereby the Earth's magnetic poles switch polarities, appear to be associated with large impacts ..."[74] Adopting this explanation makes geomagnetic reversals a fast process.

If diamonds take millions of years to form, then YES would be invalidated.   LifeGem of Chicago creates "memorial diamonds" from the carbon remains of the deceased after cremation.[75] This process just takes a matter of months! So diamonds can be created quickly. Some claim that Niagara Falls came to be over multiple thousands of years. W. J. McGee, of the U.S. Geological Survey, stated in 1893 that the Niagara gorge may have formed as rapidly as just 5,000 years.[76]

Raymond Arthur Lyttleton FRS (d. 1995) was a British mathematician and Professor of Theoretical Astronomy at Cambridge. Lyttleton concludes that due to tidal friction in the Earth-Moon system, the Moon would nearly touch the Earth just one billion years ago!! Does that mean that Lyttleton supported YES (1B < 4.5B)? Maybe not, but this is surely a certain refutation of Old Earth Fallacies (OEF's).[77] George Darwin, son of Charles, was Professor of Astronomy at the University of Cambridge and studied terrestrial tidal friction and the moon. In a 2012 journal article, Sung-Ho Na of the Korea Astronomy and Space Science Institute freely admitted that,

> Assuming a constant tidal phase lag during the whole geological past, it was found that the Moon should have been located near the Roche limit around 1.7 billion years ago. Since the Moon was formed 4.5 billion years ago, the discrepancy between the two - 1.7 and 4.5 billion years cannot be explained.[78]

a.f. wikipedia

Suppose the Moon did get too close to our lovely planet. Within the Roche limit the Moon's own gravity can no longer withstand the tidal forces and the Moon disintegrates.[79] This is no problem from the YES viewpoint, since the Earth and Moon are both young and the Moon was never that close.

In *The Earth's Variable Rotation*, Kurt Lambeck observes,

> Earth is less ephemeral than the ocean basins, but unless the present estimates for the accelerations are vastly in error, <u>the only available energy sink that can solve the time-scale problem and the only energy sink that can vary significantly with time is the ocean.</u> All the evidence now points to the fact that the oceans are a much more important energy sink. The actual mechanics of the dissipation remain unclear nevertheless. If, as generally supposed since the work of G.I. Taylor and Jeffreys, <u>dissipation is by friction in shallow seas, then any extrapolation into the past becomes very uncertain indeed since we know that important changes in the ocean configuration have occurred in the past.</u>[80]

If there was a global ocean, that would fix the problem - less friction! A global catastrophe of this nature would imply a shorter geologic timescale and favors YES. Pioneer geologist Abraham Werner proposed a global ocean. So, the Moon's tidal friction supports Global Ocean Theory (GOT). Even now, when viewed from the right angle, this world is mostly water (Australia is on the left):

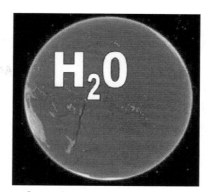

a.f. explorers.neaq.org

Katja Fennel et al describe what has been called the Oxygen Conundrum, "Geochemical evidence suggests that there was a delay of several hundred million years between the evolution of oxygenic photosynthesis and the accumulation of oxygen in Earth's atmosphere."[81] If the Earth is only thousands of years old, there is no conundrum.

William Stansfield, Emeritus Professor of Biological Sciences at CalPoly - San Luis Obispo, is at least willing to interact with evidence for YES:

> Uranium salts presently appear to be accumulating in the oceans at about one hundred times the rate of their loss. … Under uniformitarian rules, the total concentration of uranium salts of the oceans … could be accumulated in less than one million years.[82]

An astounding revelation appeared in a report in *New Scientist*. A laboratory experiment (c.1964) conducted at the Central Radio Propagation Laboratory of the US National Bureau of Standards led to a frank claim by Dr. E. E. Ferguson:

> … "some catastrophic event" may have made most of the helium boil off in geologically recent times. At first sight there ought to be about a thousand times as much helium in the atmosphere as there is. … At present known rates, radioactivity would produce all the neutral helium gas now in the atmosphere in no more than a few million years.[83]

Stansfield reveals an acute problem with radiocarbon dating:

> Since the amount of C^{14} is now increasing in the atmosphere, it may be assumed that the quantity of C^{14} was even lower in the past than at present. ... [some argue that] <u>since C^{14} has not yet reached its equilibrium rate, the age of the atmosphere must be less than 20,000 years old</u>.[84]

Keep in mind that Stansfield is a committed evolutionist and is going for the Billions. At least he is willing to dialog regarding YES.

According to Claudia Stolle of Denmark's National Space Institute, "At the moment, the Earth's magnetic field is decreasing by approximately 5 % per century, and scientists are unable to explain the reason for this or describe the consequences this will have"[85] Does the rapid decay of earth's magnetic field point to its youth?

Charles Carrigan and David Gubbins, writing in *Scientific American*, diagnose the Earth's magnetic field:

> The field cannot be descended from a magnetic field of the primordial earth because the electric conductivity of the core is too low. <u>Without a constant supply of energy the electric currents maintaining the field would have died out in less than 10,000 years</u>. And the field has clearly existed for a much longer time.[86]

This strong evidence supports YES! Mercury's *live* magnetic field also supports YES. According to NASA, "Researchers have been puzzled by Mercury's field because its iron core was supposed to have cooled long ago and stopped generating magnetism."[87] If Mercury is young, the Earth may be young as well. James Hogan notes that thermoluminescence dating of lunar material was dated at less than 10,000 years.[88]

Scottish mathematical physicist Peter Guthrie Tait (d. 1901) estimated that the Sun was 10M years old.[89] This was, of course, before we discovered that solar power's source is nuclear fusion. However, this shows that Geology could provide no obvious evidence that this world must be at least 100M years old, which was a common estimate in the late 1800's. Earth scientists' old earth bias was unfounded.

Leading Scientists Estimate Earth's Age

Alfred Russel Wallace (FRS, d. 1913), who co-published a paper on evolution with Charles Darwin in 1858, estimated that the time since the beginning of the Cambrian was only 24M years![90]

a.f. hnhs.org

Lord Kelvin's last estimate for the age of the Earth (1897) was 24M years.[91] Even as late as 1919, there were a significant number of scientists who held that the Earth was as young as 10M years.[92]

a.f. gettyimages.com

Arthur Holmes, the most important person to make radiometric methods the dominant dating method, said in 1913, "It is now well known that if the proportion of radium in the interior of the earth is the same as that in the surface rocks, the earth ought to be growing hotter..."[93] Clearly, this is not the case. This is a problem for an old earth, but not for a young earth.

The Earth is Younger than you Think

Aleksey Smirnov (Michigan Technological University) and fellow researchers (University of Rochester, Yale) have found that the Earth's inner core is likely 1.2B years older than previously thought.[94] If the core's date has been 1.2B years incorrect, maybe the age of the Earth itself is in error and it's only really thousands of years old.

William Corliss, the man behind the Sourcebook Project, posits a wise question: "Geology allied with the hypothesis of evolution has given us a multi-billion-year tapestry on which to paint the earth's history. Has science's loom gone awry?"[95] Establishment science has missed the mark by expanding the earth's age by several orders of magnitude.

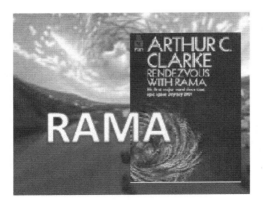

Physicist William Corliss (d. 2011) was a compiler of scientific anomalies and founder of the Science-Frontiers.com website. Corliss started the Sourcebook Project in 1974. Esteemed scifi author Arthur C. Clarke (d. 2008) said, "Corliss selected his material almost exclusively from scientific journals like *Nature* and *Science* ... there is much that is quite baffling in some of these reports from highly reputable sources." Corliss wrote 13 educational books for NASA.[96]

Darwin frankly admitted in a letter to Charles Lyell,

> I rejoice profoundly; for thinking of the many cases of <u>men pursuing an illusion for years</u>, often and <u>often a cold shudder has run through me</u> and I have <u>asked myself whether I may not have devoted my life to a phantasy</u>. Now I look at it as morally impossible that investigators of truth like you and Hooker can be wholly wrong; and therefore I feel that I may rest in peace.[97]

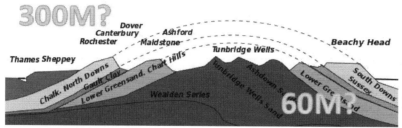

a.f. wikipedia

Darwin discussed the "denudation of the Weald," that is the time it took to erode chalk and other beds in southern England. Darwin said 300M years, whereas the standard geologic estimate is 60M years. Darwin was off by a factor of five![98] Darwin was wrong about evolution (Ch. 3) and about the age of the earth. Could we have been wrong about radiometric dating as a gauge for this planet's youth? Stop the illusion – tell the truth!! Make Big Science accountable and expose OEF's. Practice independent and critical thinking and reject the bias of Mass Geo regarding the age of the world. Defenders of billions of years have failed. In this Age of Catastrophism, the voice of Geology has no speech refuting YES.

Notes:

1) Plato, *Laws,* Book 3,
http://classics.mit.edu/Plato/laws.3.iii.html

2) Quoted in *Time's Arrow, Time's Cycle* by Stephen Jay Gould (Harvard Univ. Press, Cambridge, MA, 1987), p. 45.

3) *Kicking the Sacred Cow* by James Hogan (Baen, New York, NY, 2004), p. 175, Hogan is endorsing Velikovsky's opinion.

4) Quoted in *The Age Of The World* by Francis Haber (Johns Hopkins Press, Baltimore, MD, 1959), p. 271, emphasis added.

5) Quoted in *The Age Of The World* by Francis Haber (Johns Hopkins Press, Baltimore, MD, 1959), p. 272, emphasis added.

6) "Geology" topic in *New International Encyclopedia* (Dodd, Mead and Co., NYC,1915), pp. 595, 596.

7) *Dialogues Concerning Natural Religion* by David Hume (William Blackwood & Sons, Edinburgh, 1907), p.88, All Caps modified and emphasis added.
http://books.google.com/books?id=BGoRAAAAYAAJ

8) *Kicking the Sacred Cow* by James Hogan (Baen, New York, NY, 2004), p. 47.

9) "On the Nature of the Universe"
http://books.google.com/books?id=34ID71tG6i8C
p. 137.

10) "Geology as an Historical Science: Its Perception within Science and the Education System" by Jeff Dodick and Nir Orion, *Science & Education* 12 (2003): 197–211, p. 198, emphasis added.
http://stwww.weizmann.ac.il/g-
earth/articles/whole_articles/Science_and_Education.pdf

11) *The Earth's Age And Geochronology* by Derek York and Ronald Farquhar (Pergamon Press, Oxford, 1972), pp. 150, 151, emphasis added.

12) "What Color Were Dinosaurs?" by James Foley,
http://www.natureworldnews.com/articles/1649/20130429/what-color-
dinosaurs-test-ancient-skin-sample-will-reveal-final.htm

13) "Life in the old fossil yet," *The Economist*, Nov. 30, 2013.

http://www.economist.com/news/science-and-technology/21590874-how-remnants-dinosaur-tissue-have-survived-millions-years-life?frsc=dg|a

14) "Oldest Complex Organic Molecules Found In Ancient Fossils" by Pam Frost Gorder

http://researchnews.osu.edu/archive/foscolor.htm

15) "Proteins in the fossil bone of the dinosaur, Seismosaurus" by L.R. Gurley, J.G. Valdez, W.D. Spall, B.F. Smith, D.D. Gillette, *Journal of Protein Chemistry* 1991;10(1):75-90, p. 75.

http://www.ncbi.nlm.nih.gov/pubmed/2054066

16) "Dinosaur Shocker" by Helen Fields, *Smithsonian*, May 2006,

http://www.smithsonianmag.com/science-nature/dinosaur.html

17) "Twisted structure preserved dinosaur proteins" by Ed Yong

http://www.nature.com/news/2011/110614/full/news.2011.369.html

18) "DNA Sequences from a Fossil Termite in Ollgo-Miocene Amber and Their Phylogenetic Implications" by Rob DeSalle, John Gatesy, Ward Wheeler and David Grimaldi, *Science*, New Series, Vol. 257, No. 5078 (Sep. 25, 1992), pp. 1933-1936, p. 1933.

19) "Silurian brachiopods with soft-tissue preservation" by Mark Sutton, Derek Briggs, David Siveter and Derek Siveter, *Nature*, Vol. 436, pp. 1013-1015 (18 August 2005) .

http://www.nature.com/nature/journal/v436/n7053/abs/nature03846.html

20) "Collagen survival and its use for species identification in Holocene-lower Pleistocene bone fragments from British archaeological and paleontological sites" by Mike Buckley and Matthew James Collins, *Antiqua*, Vol. 1, No. 1 (2011).

http://www.pagepress.org/journals/index.php/antiqua/article/view/antiqua.2011.e1

21) *Petrified Wood* by Frank Daniels (Western Colorado Pub., Grand Junction, CO, 1998), p. 11.

22) *Organic Geochemistry* by Geoffrey Eglinton and Mary T. J. Murphy (Springer-Verlag, 1969), p. 459, emphasis added.

23) Ibid., p. 450.

24) "Racemization of Amino Acids in Marine Sediments" by John Wehmiller and P. E. Hare, *Science*, Sep. 3, 1971, Vol. 173, no. 4000, pp. 907-911, p. 907, emphasis added.

25) Personal communication from Mike Viney, editor The Virtual Petrified Wood Museum,

http://petrifiedwoodmuseum.org/SpecimenRidge.htm

26) "Gilboa Fossil Forest"

http://en.wikipedia.org/wiki/Gilboa_Fossil_Forest

27) "Fitzroya"

http://en.wikipedia.org/wiki/Fitzroya

28) "How Old Is That Tree?" ed. by Christopher J. Earle

http://www.conifers.org/topics/oldest.htm

Accessed 3/13/2008.

29) "Lagarostrobos"

http://en.wikipedia.org/wiki/Lagarostrobos

30) "Neanderthal"
http://en.wikipedia.org/wiki/Neanderthal

31) "Population growth"
http://en.wikipedia.org/wiki/Population_growth

32) quoted in "The Riddle of Lost Civilization – The Case for Advanced Neanderthals" by Martin Ruggles, *Atlantis Rising*, Sep./Oct. 2013, No. 101, pp. 42, 43, 69-71, p. 70.

33) "Neanderthal Burials" by Katy Meyers
http://bonesdontlie.wordpress.com/2011/04/25/neanderthal-burials/

34) "Modeling the recent common ancestry of all living humans" by Douglas Rohde, Steve Olson and Joseph Chang, *Nature*, Vol. 431,Sep. 30, 2004, p. 562, emphasis added.

35) "Estimate of the Mutation Rate per Nucleotide in Humans" by Michael Nachman and Susan Crowell, *Genetics*, Sep. 1, 2000,Vol. 156,No. 1, pp. 297-304, p. 304, emphasis added.
http://www.genetics.org/content/156/1/297.full

36) "Calibrating the Mitochondrial Clock" by Ann Gibbons, *Science*, 279: 28-29, 1998, emphasis added.
http://www.dnai.org/teacherguide/pdf/reference_romanovs.pdf

37) "Thor Heyerdahl"
http://en.wikipedia.org/wiki/Thor_Heyerdahl

38) "Ra (1972 film)"
http://en.wikipedia.org/wiki/Ra_%281972_film%29

39) "America's early people may have come by boat from Asia" by Richard L. Hill (*The Oregonian*)
http://www.latinamericanstudies.org/ancient/coastal-route.htm

40) *Maps of the Ancient Sea Kings* by Charles Hapgood (Adventures Unlimited Press, Kempton, IL, 1996), orig. ed. 1966, p. 241.

41) Ibid., pp. 79-93.

42) "Act II: The Second Voyage"
http://libweb5.princeton.edu/visual_materials/maps/websites/pacific/cook2/cook2.html

43) *Maps of the Ancient Sea Kings* by Charles Hapgood (Adventures Unlimited Press, Kempton, IL, 1996), orig. ed. 1966, p. 152.

44) Ibid., pp. 171, 176, 177.

45) "Endorsements for Secret Maps of the Ancient World"
http://www.asiaticfathers.com/endorse.htm

46) *Secret Maps of the Ancient World* by Charlotte Harris Rees, use the "Look Inside" feature,
http://www.amazon.com/Secret-Ancient-World-Charlotte-Harris/dp/1434392783

47) "Excerpt from *The Asiatic Fathers of America*" by Hendon Harris, *US–China Review*, Summer 2011, Vol. 35, No.3
http://www.uscpfa.org/USCR/USCPFA%202011%20Summer.pdf

48) "The Truth Endures" by Charlotte Harris Rees, p. 8,
"3,000-year-old Chinese Coins in Canadian Creek," p. 9,

"Asiatic Echoes: The Identification of Chinese Pictograms in North American Native Rock Writing" by John Ruskamp, p. 11, *US–China Review*, Summer 2011, Vol. 35, No.3.

49) *Lost Discoveries* by Dick Teresi (Simon & Schuster, NYC, 2002), p. 261.

50) "The age of the Earth and the invention of geological time" by J.D. Burchfield in *Lyell: the Past is the Key to the Present* ed. by D.J. Blundell and A.C. Scott, Geological Society, London, Special Publications, 143 (1998): 137-143, p. 137, emphasis added.
http://sp.lyellcollection.org/content/143/1/137.refs

51) *A History of Ancient and Early Medieval India* by Upinder Singh (Dorling Kindersley, Delhi, India, 2008), pp. 111, 144.

52) "King List of India"
http://en.wikipedia.org/wiki/List_of_Indian_monarchs

53) *Ancient Chinese Inventions* by Deng Yinke (Cambridge University Press, 2010), p. 156.

54) "Babylon"
https://en.wikipedia.org/wiki/History_of_Babylon

55) "Short chronology timeline"
http://en.wikipedia.org/wiki/Short_chronology_timeline

56) *Metaphysical Inquiry into the Method Objects and Results of Ancient and Modern Philosophy* by Isaac Preston Cory (William Pickering, London, 1833), p. 10.

57) "New Chronology (Rohl)"
http://en.wikipedia.org/wiki/New_Chronology_(Rohl)

58) "Sumer"
http://en.wikipedia.org/wiki/Sumer

59) "Awan dynasty"
http://en.wikipedia.org/wiki/Awan_dynasty

60) "List of Assyrian Kings"
http://en.wikipedia.org/wiki/List_of_Assyrian_kings

61) "Gilgamesh"
http://en.wikipedia.org/wiki/Gilgamesh

62) "Gilgamesh" by Joshua Mark
http://www.ancient.eu.com/gilgamesh/

63) "Babylonian Map of the World"
http://www.ancient.eu.com/image/526/

64) "Babylonian Culture and Tablets"
http://www2.econ.iastate.edu/classes/econ355/choi/bab.htm

65) "Sumer"
https://en.wikipedia.org/wiki/Sumer

66) "Sumerian King List"
http://en.wikipedia.org/wiki/Sumerian_king_list

67) *Shattering the Myths of Darwinism* by Richard Milton (Park Street Press, Rochester, VT, 1992), p. 32.

68) Ibid., p. 32.

69) "Reef-Building Corals" by James Porter, *Biological Report* 82(11.73)

TR EL-82-4 (U.S. Army Corps of Engineers, 1987), p. 4.
http://www.nwrc.usgs.gov/wdb/pub/species_profiles/82_11-073.pdf

70) "Great Barrier Reef"
http://en.wikipedia.org/wiki/Great_Barrier_Reef

71) "Reef Building" by Harry Ladd, *Science*, Sep. 15, 1961, Vol. 134 no.
3481 pp. 703-715, p. 703.
http://www.sciencemag.org/content/134/3481/703.extract

72) "Rapidity of marine carbonate cementation — implications for
carbonate diagenesis and sequence stratigraphy: perspective" by Gerald
Friedman, *Sedimentary Geology*, Vol. 119, Issues 1–2, July 1998, pp. 1-4,
p. 1, emphasis added.

73) "Very rapid geomagnetic field change recorded by the partial
remagnetization of a lava flow" by Scott Bogue and Jonathan Glen
Geophysical Research Letters, Vol. 37, Issue 21, Nov. 2010,
http://onlinelibrary.wiley.com/doi/10.1029/2010GL044286/abstract
emphasis added.

74) *Quakes, Eruptions and Other Geologic Cataclysms* (*Revealing the
Earth's Hazards*, Rev. Ed.) by Jon Erickson (Checkmark Books, NYC,
2001), p. 244.

75) "LifeGem"
http://en.wikipedia.org/wiki/LifeGem

76) "Note on the 'Age of the Earth'" by W.J. McGee, *Science*, June 9,
1893, Vol. 21, no. 540, pp. 309-310, p. 309.

77) *The Earth and Its Mountains* by R.A. Lyttleton (John Wiley & Sons,
Chichester, UK, 1982), p. xv.

78) "Tidal Evolution of Lunar Orbit and Earth Rotation" by Sung-Ho Na
Journal of The Korean Astronomical Society, Vol. 45: 49-57, April 2012, p.
54.
http://dx.doi.org/10.5303/JKAS.2012.45.2.49

79) "Roche Limit"
http://en.wikipedia.org/wiki/Roche_limit

80) *The Earth's Variable Rotation: Geophysical Causes and Consequences*
by Kurt Lambeck (Cambridge University Press, 1980), p. 288, emphasis
added.

81) "The co-evolution of the nitrogen, carbon and oxygen cycles in the
Proterozoic ocean" by Katja Fennel, Mick Follows and Paul Falkowski
American Journal of Science, 2005, vol. 305, no. 6-8, pp. 526-545, p. 526,
emphasis added.

82) *The Science of Evolution* by William Stansfield (Macmillan, NYC,
1977), p. 81, emphasis added.

83) Quoted in *Unknown Earth* by William Corliss (The Source Project,
Glen Arm, MD, 1980), p. 776, emphasis added.

84) *The Science of Evolution* by William Stansfield (Macmillan, NYC,
1977), p. 83, emphasis added.

85) "The Earth's magnetic field"
http://www.space.dtu.dk/English/Research/Earths_physics_and_geodesy/M
agnetic_field.aspx

86) "The Source of the Earth's Magnetic Field" by Charles Carrigan and David Gubbins, *Scientific American*, Vol. 240, pp. 118 – 130, Feb. 1979, p. 118, emphasis added.

87) "New Discoveries at Mercury" http://science.nasa.gov/science-news/science-at-nasa/2008/03jul_mercuryupdate/

88) *Kicking the Sacred Cow* by James Hogan (Baen, New York, NY, 2004), p. 206.

89) "Calculating the age of the Earth and the Sun" by Arthur Stinner, *Physics Education*, July 2002, Vol. 37, no. 4, pp. 296-305, p. 300.

90) "Darwin and the Dilemma of Geological Time" by Joe D. Burchfield, *Isis*, Vol. 65, No. 3 (Sep. 1974), pp. 300-321, p. 317. http://www.blc.arizona.edu/courses/schaffer/449/Soft%20Inhertance/Burchfield%20-%20Darwin%20and%20Geol.%20Time.pdf

91) "The Age of the Sun and the Earth" by Florian Cajori (1908) in *Determining the Age of the Earth* (*Scientific American*, 2013), p. 86. http://dafix.uark.edu/~danielk/Darwin/AOTE_issue.pdf

92) "How Old is the World" by William McNairn (1919) in *Determining the Age of the Earth* (Scientific American, 2013), p. 94. http://dafix.uark.edu/~danielk/Darwin/AOTE_issue.pdf

93) "Radium and the Evolution of the Earth's Crust" by Arthur Holmes (1913) in *Determining the Age of the Earth* (Scientific American, 2013), p. 49. http://dafix.uark.edu/~danielk/Darwin/AOTE_issue.pdf

94) "How Old is the Earth's Inner Core?" by Dennis Walikainen http://www.mtu.edu/news/stories/2011/november/story51323.html

95) *Unknown Earth* by William Corliss (The Source Project, Glen Arm, MD, 1980), p. 758.

96) "William R. Corliss" http://en.wikipedia.org/wiki/William_R._Corliss

97) "Darwin, C. R. to Lyell, Charles" http://www.darwinproject.ac.uk/letter/entry-2543 emphasis added.

98) *Time's Arrow, Time's Cycle* by Stephen Jay Gould (Harvard Univ. Press, Cambridge, MA, 1987), p. 121.

Chapter 2
Radiometric Dating: Get a Half-Life!

Oh! Do not attack me with your watch.
A watch is always too fast or too slow.
I cannot be dictated to by a watch.
 - Jane Austen

I'm waking up to ash and dust
I wipe my brow and I sweat my rust
I'm breathing in the chemicals …
Welcome to the new age …
I'm radioactive, radioactive
 - Imagine Dragons

Writing at Discovery.com, Ian O'Neill reveals that,

> … the decay rates of radioactive elements are changing.
> This is especially mysterious as we are talking about
> elements with "constant" decay rates - these values
> aren't supposed to change. … This is the conclusion
> that researchers from Stanford and Purdue University
> have arrived at…The sun might be emitting a
> previously unknown particle [or neutrinos maybe?] that
> is meddling with the decay rates of matter. …
> researchers noticed the decay rates vary repeatedly
> every 33 days - a period of time that matches the
> rotational period of the core of the sun. The solar core
> is the source of solar neutrinos.[1]

Evidence based on tree rings indicates that the solar cycle was just 7 years in the Post-Ice Age era, in contrast to the current cycle of 11 years.[2] Could more solar activity imply more rapid radioactive decay rates in the past? If the sun affects radioactive decay rates and the solar cycle was more frequent in the past, could that imply that radioactive decay was faster in the past? This planet may actually be youthful and not billions of years old.

Earth's Age before Radiometric Dating

Charles Walcott, who discovered the Burgess Shale, estimated that the time since the Cambrian was 27.6M years ago. He used sedimentation rates to arrive at this figure.[3] In 1883, Alexander Winchell, former State Geologist of Michigan, estimated that the age of the earth at 3M years.[4] Former president of the Geological Society of London, Andrew Ramsay (d. 1891), held that the earth may be 10B years old![5]

How Radiometric Dating became King

In 1921, at a meeting of the British Association for the Advancement of Science, there was no consensus on radiometric dating and an Earth that's billions of years old. William J. Solas (Univ. of Oxford) held to an Earth no older than 100M years old.[6] In 1926, a committee of the National Research Council concluded that radiometric dating was the only reliable method to date rocks. Arthur Holmes, who led the campaign for radiometric dating, wrote 70% of the report.[7] So, it has not yet been a century since those who are going for the billions can claim victory through radioactivity.

Charles Lyell estimated that the beginning of the Cenozoic was 80 million years ago.[8] The current estimate is 65.5M. Could Lyell's number have influenced future dating of the Cenozoic? That is, the Old Earth dating came first and then the "absolute" dating methods came later! Charles Schuchert pointed out that the age of the Cambrian was nailed down *before* the advent of radiometric dating: "Reade (1893) was the first to calculate the age of the Earth since the beginning of Cambrian time on the basis of the amount of limestone, getting an age of 600 million years ... Reade's figures therefore show a rather remarkable agreement with what radioactivity teaches us now."[9]

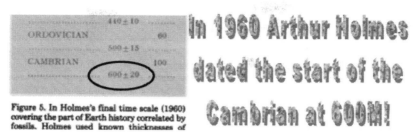

In 1960 Arthur Holmes dated the start of the Cambrian at 600M!

Figure 5. In Holmes's final time scale (1960) covering the part of Earth history correlated by fossils. Holmes used known thicknesses of

a.f. faculty.jsd.claremont.edu

J.E. O'Rourke confirms the reliance of radiometric dates on prior estimates:

> The first radiometric dates had to be calibrated by comparison with some other standard, however crude. The best previous time scales had been obtained by estimating the average rate of sedimentation... Radiometric dating would not have been feasible if the geologic column had not been erected first.[10]

Derek York and Ronald Farquhar, both from the Physics Department of the University of Toronto, conclude that, "Until the beginning of the present century [20th] geologists had no really satisfactory mechanisms for measuring time or time intervals."[11] That is, until the advent of radiometric dating, geologists' estimates for the age of the world were all over the map. But what if radioactive age metrics are terminally flawed? YES we can explore alternatives to Old Earth Fallacies (OEF's)!!

Based on the Earth's cooling, Lord Kelvin estimated that the Earth to be as young as 24M years old. Mark Twain agreed, "As Lord Kelvin is the highest authority in science now living, I think we must yield to him and accept his view."[12] Philip England, writing in *GSA Today*, reveals another Old Earth Fallacy (OEF) regarding Kelvin's age of the earth:

> The prevalent version of this tale alleges that the discovery of radioactivity simultaneously provided the demonstration (through radiometric dating) that Kelvin had greatly underestimated the age of the Earth and the explanation of why he was wrong (radioactivity being a source of heat that invalidated Kelvin's calculation). We show this popular story to be incorrect; introducing the known distribution of radioactivity into Kelvin's calculation does not invalidate its conclusion. In 1895, before the discovery of radioactivity, John Perry showed that convection in the Earth's interior would invalidate Kelvin's estimate for the age of the Earth, but Perry's analysis was neglected or forgotten …[13]

How is it that simply the existence of heat produced from radioactivity invalidated Kelvin's age of the earth of 23M years? Could it be because of the strong Old Earth bias of the scientific elites?

It is often said that stratigraphy gives us the relative dates of the rock layers, but radiometric dating gives us the absolute dates. J.E. O'Rourke corrects this OEF: "'Absolute age' is a self-contradiction. For this reason, and for giving an exaggerated impression of certainty, <u>the term was deplored by the very man who did the most to make radiometric dating practical</u> [Arthur Holmes] ..."[14]

Discordant – Miscordant (He Said/She Said)

Derek York and Ronald Farquhar freely reveal that, "The <u>majority</u> of the many published uranium-lead and thorium-lead age determinations are <u>discordant</u>."[15]

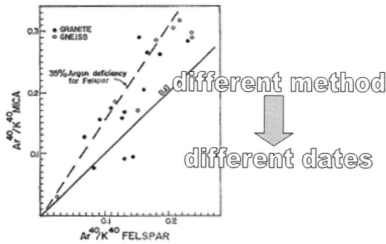

a.f. *The Earth's Age And Geochronology*

How is it that felspar and mica give different dates? Could there be a storm brewing in the halls of radiometric dating?

Location	Discordant Radiometric ages (M's of years)				
	Pb^{206}/U^{238}	Pb^{207}/U^{235}	Pb^{207}/Pb^{206}	Pb^{208}/Th^{232}	max-min
Sweedish Kolm	365	435	920		555
Uraninite, Huron Claim, Manitoba	1564	1985	2475	1273	1202
Zircon, Quartz Creek, Colorado	925	1130	1540	530	1010
Zircon, Johnny Lyon Granodiorite, Arizona	1070	1270	1630	1660	590
				Average	**839**

data from *The Earth's Age And Geochronology*

In other words, different dating methods give different ages. Could this indicate an underlying unreliability in the methods themselves? Note the difference between the high and low values: 800 million here, 80% of a Billion there and pretty soon you're talking about some real time! The Halfway House granite of South Africa was dated with the Rb-Sr method and the dates ranged from 2.3B to 4.5B depending on the mineral used.[16]

Radiometric dates are highly sensitive to proximity to igneous intrusions. Even if a rock is as far away as 10km from an igneous intrusion, the dates from hornblende, biotite and feldspar may vary by hundreds of millions of years from the established ages!![17]

Writing in the *Canadian Journal of Earth Sciences*, A. Hayatsu confesses,

> In conventional interpretation of K-Ar age data, it is common to discard ages which are substantially too high or too low compared with the rest of the group or with other available data such as the geological time scale. The discrepancies between the rejected and the accepted are arbitrarily attributed to excess or loss of argon.[18]

Hayatsu reports on the following range of dates for the North Mountain Basalt of Nova Scotia using the Potassium-Argon method:[19]

Sample	K-Ar Date
NS-44	222M
NS-45	333M
NS-46	199M
NS-47	208M
NS-52	181M

The range is an incredible 152M years – that's 28% of the time since the beginning of the Cambrian!!

Start of	1972 Est.	Current Est.	Error (M's)
Cretaceous	136	145.5	9.5
Triassic	225	251.0	26.0
Carboniferous	345	359.2	14.2
Devonian	395	416.0	21.0
Ordovician	500	488.3	11.7
Cambrian	570	542.0	28.0
		TOTAL	110.4

data from *The Earth's Age And Geochronology* and usgs.gov

If we take the beginnings of the Periods of the Geologic Time Scale from 1972 and compare with the current dates, we get a total error of 110M years!! Stansfield recognizes the trouble with radiometric dating:

> It is <u>obvious that radiometric techniques may not be the absolute dating methods that they are claimed to be</u>. Age estimates on a given geological stratum by different radiometric methods are often quite different (sometimes by hundreds of millions of years). There is no absolutely reliable long-term radiological "clock." The uncertainties inherent in radiometric dating are disturbing to geologists and evolutionists, but their overall interpretation supports the concept of a long history of geological evolution. The <u>flaws in radiometric dating methods</u> are considered by [some] to be <u>sufficient justification for denying their use as evidence against the young earth theory</u>.[20]

Some of the moon rocks returned from the Apollo missions, were dated as high as 7B and 20B years old![21]

Young Volcanics Dated Old

There are numerous cases where historical lava flows are dated at millions of years old by radiometric methods. Supposedly, isochrons help fix the problems with radiometric dating. However, Jon Davidson et al describe issues with isochrons:

The determination of <u>accurate and precise isochron ages</u> for igneous rocks requires that the initial isotope ratios of the analyzed minerals are identical at the time of eruption or emplacement. Studies of <u>young volcanic rocks</u> at the mineral scale have shown this assumption to be invalid in many instances. Variations in initial isotope ratios can result in <u>erroneous or imprecise ages</u>. …it is possible for initial isotope ratio variation to be <u>obscured in a statistically acceptable isochron</u>.[22]

If isochrons can't date young volcanic rocks, how can it date rocks thought to be millions of years old?

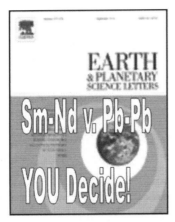

a.f. journals.elsevier.com

Catherine Chauvel et al published a revealing paper in *Earth and Planetary Science Letters*, "The Sm-Nd age of Kambalda [Australia] volcanics is 500 Ma too old!" Chauvel et al admit that, "Although Sm-Nd isotopic analyses of volcanics from Kambalda, Western Australia, form a 3.2 Ga [3.2 B] linear array, Pb–Pb analyses of the same suite give an isochron age of 2.73±0.03 Ga …"[23] Five hundred million years is 92% of the duration of the entire rock record since the time trilobites roamed the seas (Cambrian)! Cleary, this is an epic fail.

When using the K-Ar method for recent volcanic rocks, mainstream geologists often fall back on the crutch of "excess Argon" to explain away dates that are way too old. A.V. Ivanov et al make this startling acknowledgement, "... there are

<u>no strict criteria to reveal presence of extraneous argon</u> by the K-Ar method. So, it remains the <u>major limiting factor for dating young volcanic rocks</u>."[24]

Deep ocean basalts from recent lava flows from Kilauea have been dated up to 22M years old using radiometric methods![25] John Funkhouser readily admits,

> Isotopic studies have been made of the inert gases present in ultramafic xenoliths from <u>two sites in Hawaii, the 1800–1801 Kaupulehu flow (Hualalai Volcano, Hawaii) and Salt Lake Crater (Oahu). Apparent ages calculated from the measurement of radiogenic argon and helium have very high values.</u> The ratio of radiogenic helium to argon relative to natural gases and to the value expected from generation in situ is low. … The inert gas 'ages' then, are of <u>uncertain significance</u> for these samples.[26]

So when we date rocks of known age, radiometric dating fails epically. Now why should we trust this method for formations of supposedly millions of years old? Would you want someone who failed the CPA exam five times to be in charge of handling your taxes for your billion dollar company?

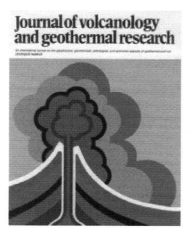

In a paper in the *Journal of Volcanology and Geothermal Research* the long ages of historic lava flows from Mount Vesuvius using the ^{230}Th dating method are explained away by "the 'contaminating' component [that] was added at subcrustal levels." Capaldi Giuseppe et al conclude, "The consequences of the <u>open-system behavior</u> of magma-forming processes with respect to Ra, U and Th <u>affect both the validity of the ^{280}Th dating method and evaluations of Th and U concentrations in the mantle.</u>"[27] Known historic deposits give wrong dates. Should we buy the dates that radiometric age measurements are selling for unknown ages (Cambrian, Ordovician etc.)?

Can We Date History?

Sheridan Bowman (British Museum Research Lab, London) confesses the truth about radiocarbon:

> A radiocarbon laboratory will also ask, what is the expected age of the sample. This is not cheating! ...samples of substantial age [10K years] must not follow modern ones. ... in the period 800-400 BC, the calibration curve is effectively flat ... and all calendar events in this period will produce approximately the same radiocarbon age.[28]

Radiocarbon can date historical objects with definite dates. Yet even that method is not accurate past 400 BC! You have to tell what the date should be to have your sample analyzed!! What's up with that? When you go to the butcher and have your meat weighed, you don't tell the butcher the weight – he/she tells you. Even Radiocarbon dating is just semi-reliable, so why should we trust the other radiometric methods that we cannot verify through historical dates?

The Radioactive Decay Constant Isn't

H.C. Dudley stated prophetically in 1976,

> ... induced changes in the disintegration rates of 14 radionuclides [have been investigated], including ^{14}C, ^{60}Co, and ^{137}Cs. The observed variations in the decay rates... were produced by changes in pressure, temperature, chemical state, electric potential, stress on monomolecular layers... These findings, together with complex and alternative disintegration modes, by which certain radio nuclides decay ([e.g.] ^{64}Cu, $ß^-$, $ß^+$ or electron capture) have led to the conclusion that ... The decay "constant" is now considered to be a variable.[29]

The half-life of ^{146}Sm (Samarium) has recently been changed from 103 ± 5 million years to 68 ± 7 million years.[30] That's a 34% reduction! Cavitation involves the forming of vacuums, which causes a plucking action on rock, due to fast moving water. Ultrasonic cavitation of water increases thorium decay by a factor of ten thousand![31]

John Joly (FRS, d. 1933), who developed radiotherapy for the treatment of cancer, was a pioneer in radiometric dating.[32] Ernest Rutherford and Joly dated Devonian granites using radiohalos. The dates ranged from 50M to 470M years. Establishment Science holds that the Devonian lasted from 416M to 359M years ago. Could this evidence have planted doubts in Joly's mind about the fundamental PreSuppostions of the radiometric dating method? As Martin Gorst notes, "... Joly stubbornly refused to accept that dates based on radioactive decay were accurate. He argued that radioactive decay had proceeded faster in the past than at present, and clung tenaciously to his estimates based on the salinity of the oceans."[33] John Joly, Chair of Geology and Mineralogy at Trinity College – Dublin, as early as 1923 recognized the problems with radiometric dating:

> We do not appear to be in a position to deny the
> possibility that uranium may have slowed down in its
> rate of decay over geological time. … The age indicated
> by uranium for Lower or Pre-Paleozoic rocks is about
> four times too great as compared with the age indicated
> by thorium. The complete tale is not yet told, but I
> think the balance of probability is in favor of an age
> between 150 and 200 millions of year for the earliest
> advent of geological conditions upon the globe.[34]

Thus, as late as 1923 a prominent leader in science was
satisfied with an earth as young as 150M years – a mere 3% of
the current estimate!!

Paul Renne et al wrote an interesting paper in *Science*,
"Absolute Ages Aren't Exactly."   There are significant
uncertainties in the decay rates of various isotopes.   In fact,
geochronologists and nuclear physicists often use different
decay rates![35]

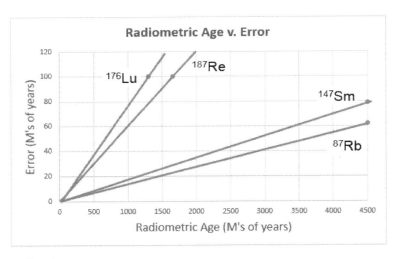

a.f. sciencemag.org

Some scientists, such as H.G. Owen, have postulated that the
Earth has expanded.   In conjunction with this they have
proposed a change in the universal gravitational constant.
Could this have affected radioactive decay rates?[36] In order to
deal with the issue of tidal friction in the Earth-Moon system,

R.A. Lyttleton thought that the universal gravitational constant (G) may vary.[37]

Even Derek Ager (*The New Catastrophism*) agrees that some of the fundamental constants have changed:

> There is … the question of die so-called gravity constant, which geophysicists tell me is immutable. I can only say that, in spite of Newton and all that: "I have heard that story before;" I remember the time when the poor innumerate geologists were enthusiastic about continental drift, but the geophysicists said that it was impossible …So far as the gravity constant is concerned, a variable one, for whatever reason, could mean a very different world in the past, including one that was much smaller than it is today.[38]

David Kitts agrees with Ager that the gravity constant may not be immutable: "[The] rate of glacial fluctuation does not remain constant throughout time any more than the force of attraction between objects remains constant throughout time."[39] Roman Tomaschitz of the Physics Department of Hiroshima University also proposes variation in the gravitational constant.[40] His motivation is to find a solution to the Faint Young Sun Paradox. According to the standard solar model, the sun billions of years ago was not as bright and a cold Earth could not support life. This is a problem that Carl Sagan had a keen interest in. If the universal gravitational constant can change, why not radiometric decay rates as well?

Derek York and Ronald Farquhar, both from the Physics Department of the University of Toronto, pointedly confess to,

> … the possibility that the various "constants of nature" (i.e. the gravitational constant G, the radioactive decay constants, etc.) may be a function of time. This matter has been considered by Dirac (1939), Dicke (1959) and Kanasewich and Savage (1963), among others.[41]

An international team of astrophysicists theorized that the basic laws of nature may be changing. Their astronomical observations on patterns of light absorption that the team could not explain without assuming a change in a basic constant of nature involving the strength of the attraction between electrically charged particles.[42] Maybe other constants regarded as immutable may also change. How could this impact radioactive decay rates? If the radioactive decay rate was much faster in the past, formation allegedly millions of years old may be only thousands. Thus, we have a greatly reduced geologic time scale.

Paul Dirac (d. 1984) predicted the existence of antimatter and won the Nobel Prize in Physics for 1933 along with Erwin Schrödinger. Dirac wrote a paper suggesting that certain fundamental constants of physics may actually change and said, "... a thorough application of our present ideas would require us to have <u>the rate of radioactive decay varying with the epoch and greater in the distant past than it is now.</u>"[43] Two Israeli physicists have discovered how an unstable nucleus might decay faster. This involves the Zeno effect and quantum tunneling. Gershon Kurizki and Abraham Kofman, both of the Weizmann Institute, published this result in *Nature*.[44] A.S. Barabash of the Institute of Theoretical and Experimental Physics of Moscow, presents a shocking idea,

> ... geochemical measurements on <u>young minerals give lower values</u> of $T_{1/2}$ (^{130}Te) [half-life of Tellurium-130] <u>as compared to measurements on old minerals</u> [in the geologic record]. It is proposed that <u>this could be due to a change in the weak interaction constant with time.</u>[45]

That is, radioactive decay rates DO change with time and one possible mechanism is via the weak interaction constant.

If radiometric dating is wrong, since decay rates are **not** constant, this orb may not be billions or millions of years old. How about going for the thousands? In a paper in the

American Journal of Science, J.E. O'Rourke comments on one possible cause of decay rate change:

> One cosmogonist, noting that there can be no preferential frames of reference in the universe, proposes the expanding galaxies as points in such a framework, which would be in effect a hyperbolic space with a time scale that takes account of the interval elapsed since the creation... On this scale, <u>spectral frequencies vary with the date, and the disintegration constant [radioactive half-lives] would not be constant</u> ...[46]

Radioactive decay rates are **not** constant. Let's look at a further example. Under a fully ionized state, [187]Re can decay <u>billions of times faster</u> than its normal rate. The [187]Re half-life went from 42B years to just 33 years!![47] These facts are clearly consistent with Young Earth Science (YES).

Got Groundwater?

Douglas Keenan describes problems with radiocarbon dating,

> Along the coast of Ecuador, a study of the <u>surface atmosphere ^{14}C age during 1992–1993 found that the age abruptly increased by over 350 ^{14}C years</u> immediately after the onset of the upwelling of old Pacific water (related to El Niño) ...In northwestern Thailand, a study of tree rings grown during 1952–1975 found that the ^{14}C ages of rings from 1953 and 1954 were <u>roughly 200 ^{14}C years too old</u> ...The investigators suggested that this was due to exceptional monsoonal <u>upwelling of very old water</u> in the northern Indian Ocean. The location where the trees grew is about 225 km inland.[48]

Could groundwater effects have influenced other dating methods (K-Ar, U-Pb etc.)? M.R. Klepper and D.G. Wyant endorse this possibility:

Most igneous rocks also contain uranium in a form that is readily soluble in weak acids. … as much as 90 percent of the total radioactive elements of some granites could be removed by leaching … significant quantities of uranium can be leached from igneous rocks by acidic ground water …[49]

According to Trausti Einarsson of the University of Iceland,

In radiometric dating there are several source of error which seem to have been given less attention than is desirable, and which are discussed in this paper. The main topics are: indications of the effects of groundwater circulation of the loss of Ar and Sr… Besides absorption of radon, the many effects of groundwater reaching depths with temperatures of 100-150 °C include at least partial dissolution of such minerals as feldspars … It seems hardly possible to assume that rocks of an age above 50-100 million years have escaped such effects of groundwater …[50]

Cheng Hung's alternative dynamic simulation model for radiometric dating yields this revealing example:

The results indicate that the age of the rhyolite is on the order of 11,300 to 11,900 years, as compared to the age of 1,030,000 years reported in Getty and Depaolo's study. This huge discrepancy in the Alder Creek rhyolite chronology indicates that it is necessary to reevaluate the accuracy of the conventional model, which utilizes a closed system assumption [and ignores the effects of groundwater].[51]

Here Comes the Sun

Periodic fluctuations in a number of radioactive decay rates correlate with the Earth-Sun distance. One possible cause involves the influence of neutrinos. Another potential explanation supposes a seasonal variation in fundamental constants.[52]   Jere Jenkins and Ephraim Fischbach (Physics Dept., Purdue) et al explain:

> ...recent work by Barrow and Shaw provides an example of a type of theory in which the Sun could affect both the alpha- and beta-decay rates of terrestrial nuclei. In their theory, the Sun produces a scalar field - which would modulate the terrestrial value of the electromagnetic fine structure constant α_{EM}. This could, among other effects, lead to a seasonal variation in alpha and beta decay rates, both of which are sensitive to α_{EM}.[53]

The Economist describes amazing research on radioactivity:

> Jere Jenkins and Ephraim Fischbach, from Purdue University ... report in *Astroparticle Physics*, the decay rate of chlorine-36 increases as Earth approaches the sun. ...Such discrepancies might be explained if a neutrino somehow amplifies the decay rates. ...if solar neutrinos transferred a mere millionth of their energy to a decaying nucleus, that might have a big effect on the rate at which it breaks up. ...Fischbach admits that while whatever process generated the flare in 2006 also caused a dip in neutrino flux, and a corresponding drop in radioactive decay rates, other processes seem to have the opposite effect. For example, a storm in 2008 was preceded by a spike in manganese-54 decay rates.[54]

Are Radiometric Ages Accurate?

Writing in the journal *Geology*, Ajoy Baksi provides this warning about relying on radiometric dates: "Subjective, and in many instances, incorrect use of radiometric data has become endemic in the earth science literature. Mathematical analysis of imperfect, and in many cases, highly subjective data sets lead to dubious conclusions."[55]

a.f. wikipedia

One instance of dating confusion involves the Devils Postpile near Mammoth Mountain in California has been dated from 100K to 700K by various methods – why the discrepancy?[56] Could there be something seriously wrong with radiometric dating? According to a web page of the U.S. Geological Survey (USGS),

> …the rocks of the Devils Postpile formed during the ice age… Less than 100,000 years ago, basalt lava, which was to become the Devils Postpile, erupted in the already glaciated valley of the Middle Fork of the San Joaquin River. The age of volcanic rocks can be estimated by study of the radioactive decay of elements in the rocks. <u>Previous estimates for the age of the Postpile basalt, ranging from about 600,000 years to nearly a million years</u>, are now thought to be seriously in error. Although an exact age for the Postpile flow still is not known, we believe that <u>an age of less than 100,000 years</u>, based on radiometric age determinations on rocks thought to be correlative, is more reasonable.[57]

From 1M to 100K, that's a ten-fold age reduction!

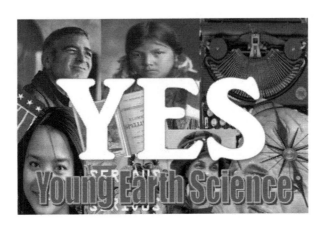

Carbon Dating says YES!

David Lowe of the National Center for Atmospheric Research in a piece in *Radiocarbon* admits that,

> Many ^{14}C dating laboratories have established that coal samples exhibit a finite ^{14}C age, apparently caused by contamination …Some of the background samples tested by Vogel … were specimens of anthracite coal which showed ^{14}C concentrations considerably higher than expected due to machine background ages and contamination during sample preparation. When background ages in the vicinity of 50 kyr were expected, the coal samples yielded 40-45 kyr. There are many other unpublished accounts by ^{14}C laboratories in which the use of coal as a background test material has been investigated. In many cases, the samples were found to contain ^{14}C, and further studies were discontinued.[58]

Maybe it's not contamination or background. Is it possible that the coal is actually less than 50K years old? So we challenge those who march to the beat of OEF's to go to their nearest coal mine and have a sample dated by the radiocarbon method. I predict an age of less than 50K years.

Consider a stick of butter. How long has it been in the house? If this butter had been left outside in Ecuador, how long would it last? Is there such a thing as a foot-long stick of butter? Has this family been on vacation for three weeks? At what rate is the butter being used? Does this family have eight kids? Notice that there are many PreSuppositions that affect our answer. The same is true for radiometric dating. The initial conditions and the family's rate of butter consumption are the key ingredients to date the age of the butter.

In 2012, at a geoscience conference in Singapore, Dr. Thomas Seiler (of Stuttgart, Germany) revealed that radiocarbon dating of a number of dinosaur bones dated to less than 50K years! This controversial result was presented at a meeting that was sponsored by the American Geophysical Union (AGU) and the Asia Oceania Geosciences Society (AOGS).[59]

Radiometric Dating is Not All That

In 1968, Don Eicher (Univ. of Colorado) had a wise caution regarding radiometric dating:

> Virtually all scientists who have seriously investigated the technique of radiometric dating agree that it rests on a firm theoretical basis. The numerous assumptions on which radiometric methods depend, such as the invariability of half-lives through geologic time, seem reasonable. ... Still, suspicions have persisted that, because of some unanticipated source of systemic error, the whole radiometric calendar from bottom to top might be drastically wrong.[60]

According to famed paleontologist George Gaylord Simpson (d. 1984), radiometric dating is not ready for prime-time:

> ... [It is of] major operational importance that paleontology, when applicable, has the highest resolving power of any method yet discovered for determining the sequence of strictly geological events. (That radiometric methods may give equal or greater resolution is at present a hope and not a fact).[61]

But does this not suppose that evolution actually occurred? We deal with evolution in the next chapter.

Notes:

1) "Is the Sun Emitting a Mystery Particle?" by Ian O'Neill,
http://news.discovery.com/space/do-solar-flares-change-the-nature-of-constant-radioactive-elements.htm

2) "Radiocarbon Analysis of Single-Year Tree Rings at the Last Glacial-Deglacial Transition" by M. Katsumata, K. Minoura, K. Horiuchi, Y. Sibata, *American Geophysical Union*, Fall Meeting 2007, abstract #PP33B-1289
http://adsabs.harvard.edu/abs/2007AGUFMPP33B1289K

3) *The Age of the Earth* by George Becker (1936)
http://www.archive.org/stream/ageearth00beckgoog/ageearth00beckgoog_djvu.txt

4) "Celebrating the age of the Earth" by Simon Knell and Cherry Lewis, *Geological Society, London, Special Publications*, 2001, vol. 190, p. 1-14, p. 7.

5) "John Perry's neglected critique of Kelvin's age for the Earth" by Philip England, *GSA Today*, Vol. 17, No. 1, Jan. 2007, p. 5.
ftp://rock.geosociety.org/pub/GSAToday/gt0701.pdf

6) "The Age of the Earth Debate" by Lawrence Badash, *Scientific American*, Aug. 1989, pp. 90-96, p. 93, 96.

7) Ibid., p. 96.

8) *Understanding Paleontology* by P.R. Yadav (Discovery Pub. House, New Delhi, 2009), p. 133.

9) "Geochronology or The Age of the Earth on the Basis of Sediments and Life" by Charles Schuchert, *Bulletin of the National Research Council*, No. 80 (National Research Council, Wash. DC, 1931),pp. 10-64, p. 21
http://books.google.com/books?id=OCcrAAAAYAAJ

10) "Pragmatism versus materialism in stratigraphy" by J.E. O'Rourke, *American Journal of Science*, Vol. 276 (1976), no. 1, pp. 47-55, p. 54.

11) *The Earth's Age And Geochronology* by Derek York and Ronald Farquhar (Pergamon Press, Oxford, 1972), p. 2.

12) "John Perry's neglected critique of Kelvin's age for the Earth" by Philip England, *GSA Today*, Vol. 17, No. 1, Jan. 2007, pp. 4-9, p. 8.
ftp://rock.geosociety.org/pub/GSAToday/gt0701.pdf

13) Ibid., p. 4.

14) Ibid., p. 54, emphasis added.

15) *The Earth's Age And Geochronology* by Derek York and Ronald Farquhar (Pergamon Press, Oxford, 1972), p. 85, emphasis added.

16) Ibid., p. 81

17) *Geologic Time* by Don Eicher (Prentice-Hall, Englewood Cliffs, NJ, 1968), p. 137.

18) "K-Ar isochron age of the North Mountain Basalt, Nova Scotia" by A. Hayatsu, *Canadian Journal of Earth Sciences*, Vol.16, 1979, p.974.

19) Ibid.

20) *The Science of Evolution* by William Stansfield (Macmillan, NYC, 1977), p. 84, emphasis added.

21) *Doomsday: the Science of Catastrophe* by Fred Warshofsky (Reader's Digest Press, NYC, 1977), p. 65.

22) "Mineral isochrons and isotopic fingerprinting: Pitfalls and promises" by Jon Davidson, Bruce Charlier, John Hora and Rebecca Perlroth, Geology, Vol. 33 (2005), no. 1, pp. 29-32, p. 29, emphasis added.

23) "The Sm-Nd age of Kambalda volcanics is 500 Ma too old!" by Catherine Chauvel, Bernard Dupré and George Jenner, *Earth and Planetary Science Letters*, Volume 74, Issue 4, August 1985, Pages 315–324, p. 315. http://www.sciencedirect.com/science/article/pii/S0012821X85800030

24) "Achievements and Limitations of the K-Ar and 40Ar/39Ar Methods: What's in It for Dating the Quaternary Sedimentary Deposits?" by A.V. Ivanov, A.A. Boven, S.B. Brandt, I.S. Brandt, S.V. Rasskazov, *Berliner Paläobiologische Abhandlungen*, Vol. 4 (2003), pp. 65-75, p. 68, emphasis added. http://hbar.phys.msu.ru/gorm/dating/ivanov-et-al-sial.pdf

25) "Deep-Ocean Basalts: Inert Gas Content and Uncertainties in Age Dating" by C.S. Noble and J.J. Naughton in *Unknown Earth* by William Corliss (The Source Project, Glen Arm, MD, 1980), p. 806.

26) "Radiogenic helium and argon in ultramafic inclusions from Hawaii" by Funkhouser, John G., and Naughton, John J., *Journal of Geophysical Research*, vol. 73, Issue 14, July 1968, pp.4601-4607, p. 4601, emphasis added. http://onlinelibrary.wiley.com/doi/10.1029/JB073i014p04601/abstract

27) "Th isotopes at Vesuvius: Evidence for open-system behaviour of magma-forming processes" by Capaldi Giuseppe, Cortini Massimo and PeceRaimondo, *Journal of Volcanology and Geothermal Research,* Vol. 14, Issues 3–4, Dec. 1982, pp. 247–260, p. 247, emphasis added. http://www.sciencedirect.com/science/article/pii/0377027382900646

28) *Radiocarbon Dating* by Sheridan Bowman (Univ. of California Press/British Museum, London, 1990), p. 55. http://books.google.com/books?id=SjsSugCVvC0C

29) *The Morality of Nuclear Planning* by H.C. Dudley (Kronos Press, Glassboro, NJ, 1976), p. 53, emphasis added.

30) "A Shorter ^{146}Sm Half-Life Measured and Implications for ^{146}Sm-^{142}Nd Chronology in the Solar System" by N. Kinoshita et al. http://www.sciencemag.org/content/335/6076/1614.short

31) "Ultrasonic cavitation of water speeds up thorium decay" by John Swain,
http://cerncourier.com/cws/article/cern/39158

32) "John Joly"
http://en.wikipedia.org/wiki/John_Joly

33) *Measuring Eternity* by Martin Gorst (Broadway Books, NYC, 2001), p. 206, emphasis added.

34) Quoted in *Unknown Earth* by William Corliss (The Source Project, Glen Arm, MD, 1980), p. 793, 794, emphasis added.

35) "Absolute Ages Aren't Exactly" by Paul Renne, Daniel Karner and Kenneth Ludwig, *Science*, Dec. 4, 1998, Vol. 282, no. 5395, pp. 1840, 1841.
http://www.sciencemag.org/content/282/5395/1840

36) *Catastrophism* by Richard Huggett (Verso, London, 1997), pp. 133, 135.

37) *The Earth and Its Mountains* by R.A. Lyttleton (John Wiley & Sons, Chichester, UK, 1982), p. xxii.

38) *The New Catastrophism* by Derek Ager (Cambridge University Press, 1993), pp. 166, 167.

39) "The Theory of Geology" by David B. Kitts in *The Fabric of Geology* ed. by Claude Albritton (Addison-Wesley, Reading, MA, 1963), pp. 49-68, p. 66.

40) "Faint Young Sun, Planetary Paleoclimates and Varying Fundamental Constants" by Roman Tomaschitz, *International Journal of Theoretical Physics*, February 2005, Vol. 44, Issue 2, pp. 195-218.
http://link.springer.com/article/10.1007%2Fs10773-005-1492-4

41) *The Earth's Age And Geochronology* by Derek York and Ronald Farquhar (Pergamon Press, Oxford, 1972), p. 6, emphasis added.

42) "Space data suggest cosmic laws alter as universe ages" by James Glanz and Dennis Overbye, *New York Times*, August 15, 2001,
http://www.sfgate.com/news/article/Space-data-suggest-cosmic-laws-alter-as-universe-2890398.php

43) "A new basis for cosmology" by P. A. M. Dirac, *Proceedings of the Royal Society A* (London), April 1938, vol. 165, no. 921, 199-208, p. 204, emphasis added.
http://rspa.royalsocietypublishing.org/content/165/921/199.full.pdf+html

44) "Furtive Glances Trigger Radioactive Decay" by Charles Seife,
http://news.sciencemag.org/2000/05/furtive-glances-trigger-radioactive-decay

45) "Is the weak interaction constant really constant?" by A. S. Barabash, *The European Physical Journal A*, July 2000, Vol. 8, Issue 1, pp. 137-140, p. 137, emphasis added.
http://link.springer.com/article/10.1007%2Fs100500070128

46) "Pragmatism versus materialism in stratigraphy" by J.E. O'Rourke, *American Journal of Science*, Vol. 276 (1976), no. 1, pp. 47-55, p. 53, emphasis added.

47) "Observation of Bound-State β- Decay of Fully Ionized [187]Re" by F. Bosch, T. Faestermann et al, *Physical Review Letters*, 77: 5190–5193 (1996).

48) "Why Early-Historical Radiocarbon Dates Downwind From The Mediterranean Are Too Early" by Douglas Keenan, *Radiocarbon*, Vol. 44, No. 1, 2002, pp. 225–237, p. 229, emphasis added.

49) "Notes on the Geology of Uranium" by M.R. Klepper and D.G. Wyant in *Contributions to the Geology of Uranium*, 1956 (U.S. Geological Survey Bulletin No. 1046, 1959), pp. 93, 94, emphasis added.
http://books.google.com/books?id=OqYPAAAAIAAJ

50) "Several Problems in Radiometric Dating" by Trausti Einarsson, *Jökull*, Vol. 25, pp. 15-33, p. 15, emphasis added.

51) "Isotopic Geochronology by Means of a Dynamic Simulation Model - Part 1, Uranium Series Method" by Cheng Hung,
http://adsabs.harvard.edu/abs/2004AGUFM.V51C0583H

52) "Evidence for Correlations Between Nuclear Decay Rates and Earth-Sun Distance" by Jere Jenkins, Ephraim Fischbach, John Buncher, John Gruenwald, Dennis Krause and Joshua Mattes, pp. 1-4.
http://arxiv.org/pdf/0808.3283v1.pdf

53) Ibid., p. 3.

54) "And now, the space-weather forecast"
http://www.economist.com/blogs/babbage/2012/08/neutrinos-and-solar-storms

55) Quoted in *Evolution, Creationism, And Other Modern Myths* by Vine Deloria Jr. (Fulcrum Pub., Golden, CO, 2002), p. 217.

56) "Devils Postpile National Monument"
http://en.wikipedia.org/wiki/Devils_Postpile_National_Monument

57) "Devils Postpile National Park Geologic Story," emphasis added,
http://geomaps.wr.usgs.gov/parks/depo/dpgeol4.html

58) "Problems Associated with the use of Coal as a Source of ^{14}C-Free Background Material" by David Lowe, *Radiocarbon*, Vol. 31, No. 2, 1989, pp. 117-120, p. 117.

59) "Carbon-14 dated dinosaur bones - under 40,000 years old"
http://www.youtube.com/watch?v=QbdH3l1UjPQ

60) *Geologic Time* by Don Eicher (Prentice-Hall, Englewood Cliffs, NJ, 1968), p. 139, emphasis added.

61) "Historical science" by George Gaylord Simpson in *The Fabric of Geology* ed. by Claude Albritton (Addison-Wesley, Reading, MA, 1963), p. 26, emphasis added.
https://archive.org/details/fabricofgeology007224mbp

Chapter 3
Stasis of Essential Types of Life

Science is real
From the Big Bang to DNA
Science is real
From evolution to the Milky Way
 - They Might Be Giants

Molding into a worm like state,
No fish around will mistake for bait,
Fusing together and spurting fins,
A spiky tail and sharpened grins. …
Many years pass and society takes hold,
What once were fish, now wise and old.
They close their eyes, contented and calm,
Sleeping forever and safe from harm.
 - Shaun Hellend[1]

Essentialism is Essential

In 2014, Richard Clinton Dawkins announced that the topic of this chapter deserves retirement,

> Essentialism - what I've called "the tyranny of the discontinuous mind" - stems from Plato … essentialism has been applied to living things and Ernst Mayr blamed this for humanity's late discovery of evolution - as late as the nineteenth century. If, like Aristotle, you treat all flesh-and-blood rabbits as imperfect approximations to an ideal Platonic rabbit, it won't occur to you that rabbits might have evolved from a non-rabbit ancestor …[2]

Christopher Gill in "Essentialism in Aristotle's biology" (2011) explains,

> ... Aristotle's readiness to regard both concrete individuals and general classes of individuals (types or species) as real entities (substances) and objects of knowledge provided a theoretical framework which validated empirical investigation of forms of animal life ...[3]

Denis Walsh, writing in the *British Journal of the Philosophy of Science*, makes Aristotle's view clear, "According to Aristotelian essentialism, the nature of an organism is constituted of a particular goal-directed disposition to produce an organism typical of its kind."[4] Walsh continues,

> Generations of commentators have thus imputed to Aristotle a form of "typological essentialism"... according to which there is a single canonical set of unchanging properties that defines a particular species [or rather Essential Types of Life = ETL] and these properties constitute the essence of its members.[5]

Walsh admits that, "typological essentialism [is] probably inimical to evolution."[6]

K. Schwenk and G. Wagner support stasis while trying to defend Darwin:

> On the one hand, phenotypes must be mutable and therefore responsive to the constantly changing demands of the environment. ... On the other hand, phenotypes must be stable so that the complex dynamics of their developmental and functional systems are not disrupted. ... It is this fundamental tension - between mutability and stability - that current evolutionary biology seeks to explain.[7]

Birds, fish, gators, bats, lizards, leaches, turtles and buffalo clearly belong to different ETL's.

Stephen Jay Gould (d. 2002) made this startling admission:

> As we survey the history of life since the inception of multicellular complexity in Ediacaran times … one feature stands out as most puzzling - the lack of clear order and progress through time among marine invertebrate faunas. We can tell tales of improvement for some groups, but <u>in honest moments we must admit that the history of complex life is more a story of multifarious variation about a set of basic designs</u> [Essential Types of Life, ETL's] than a saga of accumulating excellence. <u>The eyes of early trilobites, for example, have never been exceeded for complexity or acuity by later arthropods.</u> Why do we not find this expected order [from simple to complex life]?[8]

Verne Grant (University of Texas at Austin) rightly comments,

> Essentialism naturally dominated the thinking in the early history of biology. Here it took the form that Mayr … has called typological thinking. This is the view that individual organisms are the imperfect and hence variable manifestations of the archetype of the species [or rather ETL's] to which they belong.[9]

Francis Hitching pointedly remarks on the role of biological stasis:

> The main function of the genetic system, quite clearly, is one of renewal, of maintenance of the *status quo*, of establishing limits to change. Living cells duplicate themselves with near-total fidelity. … There are also built-in constraints. … Fruit flies refuse to become anything but fruit flies under any circumstances yet devised. The genetic system, as its first priority, conserves, blocks, and stabilizes.[10]

THE

AGE OF THE WORLD

BY THE LATE

J. LOGAN LOBLEY, F.G.S., F.R.G.S.

LATE PROFESSOR OF PHYSIOGRAPHY AND ASTRONOMY,
CITY OF LONDON COLLEGE.

AUTHOR OF "MOUNT VESUVIUS," "GEOLOGY FOR ALL," "THE TOUR,'
"HAMPSTEAD HILL," "GEOLOGY OF EALING,"

J. Logan Lobley (*Geology for All*) stated early in the last century,

> In the plant-world ... the persistence of types is conspicuous. The oldest land plants we know are ferns, very like recent ferns ... animal and plant life has existed with similar forms and similar physiological powers from the far back Cambrian period to our own times.[11]

Good science supports the stasis of Essential Types of Life (ETL's).

Is Evolution a FACT?

In *Darwin and the Darwinian Revolution*, Gertrude Himmelfarb reminds us of a forgotten aspect of Darwin's *Origin of Species*,

> It was probably less the weight of the facts than the weight of the argument that was impressive. The reasoning was so subtle and complex as to flatter and disarm all but the most wary intelligence. Only upon close inspection do the faults of the theory emerge. And this close inspection, by the nature of the case, was largely vouchsafed. The points were so intricately argued that to follow them at all required considerable patience and concentration ...[12]

One issue of *Discover* magazine had "Darwin on Trial" on its cover – is he guilty or innocent of scientific deception? According to Cyril Dean Darlington (*Darwin's Place in History, The Facts of Life*), Darwin,

> ... was able to put his ideas across <u>not so much because of his scientific integrity</u>, but because of his <u>opportunism</u>, his <u>equivocation</u> and his <u>lack of historical sense</u>. Though his admirers will not like to believe it, he accomplished his revolution by <u>personal weakness and strategic talent more than by scientific virtue</u>.[13]

Darlington (FRS, d. 1981) held the Sherardian chair of botany at Oxford University and discovered the mechanics of chromosomal crossover.

The November 2004 issue of *National Geographic* asked "Was Darwin Wrong?" Georges Cuvier (d. 1832), the catastrophist, would say "Yes." As pro-catastrophism author Trevor Palmer (Vice-Chancellor for Academic Development at Nottingham Trent University) explains:

> He [Cuvier] opposed the possibility of evolution ... because his anatomical studies showed that each species [or rather Essential Types of Life, ETL's] possessed a <u>high degree of structural organization, making a transition, particularly a gradual transition, from one to another extremely unlikely.</u>[14]

Mathematical biologist Sir D'Arcy Thompson (FRS, d. 1948) wrote *On Growth and Form* and held to the reality of essential types of life. Each animal has a body plan (bauplan) and intermediate forms do not exist.[15] Thompson believed in something like the hopeful monster theory, where evolution took giant leaps. Sound judgment favors the fixity of essential types of life.

According to Stephen Jay Gould, it is possible the human evolution did **not** happen:

> Wind back the tape of life to the early days of the Burgess Shale [505M]: let it play again from an identical starting point, and the chance becomes

vanishingly small that anything like human intelligence would grace the replay.[16]

Consider another challenge to evolution - the humble platypus: Where are the intermediate forms ("missing links") leading up to the platypus among its evolutionary ancestors. They have five X chromosomes and five Y chromosomes – wow, that's different! They give milk for their young in a unique manner, they lay eggs and the males' hind leg spurs have poison - no other mammal can make that claim! The beaver-tailed platypus has an electrosensitive bill which helps it hunt underwater with closed eyes.[17]

The assassination of JFK is a fact, but is microbes-to-Microshift evolution a certain fact? Roger Stone has recently written *The Man Who Killed Kennedy: The Case Against LBJ* claiming that Lyndon Johnson plotted to kill JFK (d. 11-22-63). There are dozens (hundreds?) of conspiracy theories revolving about this tragedy in Dallas that occurred within the memory of many people living today. We have video and eyewitness evidence for this event and yet no consensus. How can we say we "know" evolution happened when we have neither video nor eyewitness evidence and it happened millions of years ago? Did the U.S. national anthem come from a fish? Who knows?

Scholars Against Darwinism

Anomalist William Corliss prophetically announced in 1980 that,

This is the fabric of evolution; a tale of ascendancy that could not be told without geology. <u>Evolution, though, is only a theory, and all scientists must be psychologically prepared for its eventual refutation.</u> Not tomorrow, perhaps not for a century; but <u>evolution will make way for some better hypothesis.</u>[18]

Stansfield comments on a conference held at the Wistar Institute in Philadelphia in 1966,

> The mathematicians charged that if natural selection had to choose from the astronomically large number of available alternative systems by means of the mechanisms described in current evolution theory, the chances of producing a creature like ourselves is virtually zero. Murray Eden [MIT] was especially concerned about the element of randomness, inherent in the mutational process, which presumably provides the raw materials for the process of evolution. ... Eden contends that "No currently existing formal language can tolerate random changes in the symbol sequences which express its sentences. Meaning is almost invariably destroyed."[19]

To illustrate, here is a garbled sentence without meaning: Djadome ikelpof lei neuyt!

Alfred Russel Wallace (d. 1913) co-discovered evolution by natural selection along with Darwin and was a pioneer in evolutionary biogeography. Wallace had reservations though. He held that humanity's artistic, mathematical and musical skills could not be a product of natural selection.[20]

Stansfield takes four pages in *The Science of Evolution* to summarize Norman Macbeth's *Darwin Retried* which is highly critical of Darwnism.[21] According to Karl Popper, *Darwin Retried* is "a truly valuable book."[22] Stansfield has a very positive attitude in that he is willing to interact with his opponents. How many university biology professors will emulate Stansfield?

Pierre-Paul Grassé (d. 1985), zoologist at the University of Paris, had over 300 publications and supervised the 52-volume *Traité de Zoologie*. Grasse concludes,

> Today, our duty is to <u>destroy the myth of evolution,</u> considered as a simple, understood, and explained phenomenon which keeps rapidly unfolding before us. Biologists must be encouraged to <u>think about the</u> <u>weaknesses of the interpretations and extrapolations</u> that theoreticians put forward or lay down as established truths. The deceit is sometimes unconscious, but not always, since some people, owing to their sectarianism, purposely overlook reality and refuse to acknowledge the inadequacies and the falsity of their beliefs.[23]

At the beginning of evolution Grasse points out a BIG problem for Darwinists, "On the animal side, the link between the uni- and multicellular organisms is still missing. In spite of extensive study, the origin of the Metazoa is still unknown."[24] Where are the two-celled creatures, or the eleven-celled or even the 23-celled critters? It's an enigma. The fossil record and the current biota are silent. As Grasse observes, Darwinists are dumbfounded on the source of new biological information: "No formation of new genes has been observed by any biologist, yet without it evolution becomes inexplicable ..."[25] Grasse confirms that mistakes cannot make a man:

> A cluster of facts makes it very plain that Mendelian, allelomorphic mutation plays no part in creative evolution. It is, as it were, a more or less pathological fluctuation in the genetic code. It is an accident on the "magnetic tape" on which the primary information for the species is recorded.[26]

James Hutton, who promoted deep time, supported the fixity of Essential Types of Life (ETL's) and rejected evolution. Even Charles Lyell (*Principles of Geology*), in his early years, opposed evolution.[27] Richard Huggett provides this insight,

> Gregor Mendel's laws of inheritance were "rediscovered" in 1900 and provided, at last, the basis for a coherent theory of heredity. Curiously, the arrival of Mendelian genetics caused an initial decline in Darwinism, a situation which persisted through to the 1930's.[28]

Huggett further remarks,

> … the existence of catastrophic events capable of producing mass extinctions seems beyond doubt. If these catastrophic changes in the physical world have actually led to mass extinctions, then the neo-Darwinian system is dealt a death blow. … if mass extinctions are caused by occasional bombardment episodes [such as dinos at the KT boundary], if they are more frequent, more rapid, more extensive and more qualitatively different in effect than traditionally expected, the microevolutionary processes invoked by the neo-Darwinians are inadequate to explain the shape of the biosphere. Thus <u>it is possible that the current revival of catastrophism might well signal the downfall of Lyellian and traditional neo-Darwinian doctrines</u>.[29]

Thus, catastrophism may lead to the downfall of Darwiniac dominance.

a.f. wikipedia

In 1987, Dr. Benjamin Carson was the first surgeon to separate twins conjoined at the head. At age 32, Benjamin Carson became Director of Pediatric Neurosurgery at Johns Hopkins. In Beverly Hills in 2006 Dr. Carson was part of a panel discussion on evolution along with Richard Dawkins, Daniel Dennett and Francis Collins. Dr. Carson opposed evolution.[30] Dr. Carson received the Presidential Medal of Freedom in 2008.

Evolution v. Aristotle

If evolution fails, what alternative WorldView can replace Darwinism? Stasis of Essential Types of Life (ETL's).

David B. Kitts (*The Structure of Geology*) said in 1981,

> Aristotle, to a greater extent than almost anyone we know about, relied upon his observations. He observed that individual members of a species do not persist, but kinds do persist. That is a pretty obvious fact about the world. If there is abundant empirical support for the view that species persist, why do evolutionists suppose they do not persist? Evolutionists have a very elaborate abstract theory that compels us to suppose that species do not persist. Our reason for thinking that species do not persist is not our observation that they do not persist, but it is a theory that requires them not to persist. Whatever the observations that support

evolutionary theory are, they are not the observations that one species turns into another. Evolutionary theory compels us to see the fossil record as evidence of evolution. The paleontological record supports evolutionary theory if you presuppose evolutionary theory. It is consistent with evolutionary theory, but it does not compel us to accept evolutionary theory. The fossil record is consistent with and astronomical number of theories. The fossil record does not prove evolution; nothing proves evolution. Evolution is a scientific hypothesis. Anybody who says evolution is a proven fact is an idiot and deserves to be criticized.[31]

The Cambrian Explosion

Most of the major animal phyla (near the top of the classification hierarchy) appear abruptly in the Cambrian. The transitional forms that evolutionists would expect to find in the Pre-Cambrian are missing. This is known as the Cambrian Explosion or "Biology's Big Bang." Charles Darwin squirmed a little in the sixth edition of his magnum opus in order to explain this conundrum:

> With respect to the absence of strata rich in fossils beneath the Cambrian formation … though our continents and oceans have endured for an enormous period …formations much older than any now known may lie buried beneath the great oceans. With respect to the lapse of time not having been sufficient since our planet was consolidated for the assumed amount of organic change, and this objection, as urged by Sir William Thompson [Lord Kelvin], is probably one of the gravest as yet advanced, I can only say, firstly, that we do not know at what rate species change, as measured by years, and secondly, that many philosophers are not as yet willing to admit that we know enough of the constitution of the universe and of the interior of our globe to speculate with safety on its past duration.[32]

So, we know evolution happened, but the age of the Earth derived from physics is totally up in the air. Let's go look for the transitional forms that lead to the Cambrian Explosion (Є XPLO) under the sea – great idea!! This shows Darwin's commitment to evolution no matter what the facts were.

Stephen Jay Gould concludes that, "... the problem of the Cambrian explosion has remained as stubborn as ever ..."[33] There are twenty to thirty animal phyla in the modern era. According to Gould, the Burgess Shale of British Columbia represents 15-20 phyla. At the beginning of the Cambrian there were already **dozens** of phyla!! This is a major challenge to the reasonableness of evolution.[34] Francis Hitching (*The Neck of the Giraffe*) concisely states the enigma of the Cambrian Explosion, "... what sudden event caused the single-celled creatures to develop into highly complex multi-cellular ones [?] ..."[35]

Darwin v. Darwin

In the sixth edition of *The Origin of Species*, Darwin denied his own theory and became a Lamarkian: "... the adaptations being aided in many cases by the <u>increased use or disuse of parts,</u> being affected by the <u>direct action of external conditions of life</u> ..."[36]

Because of Lord Kelvin's and other scientists' short estimates for the age of the earth, Darwin himself betrayed his own theory and adopted catastrophism:

> [James] Croll estimates that about sixty million years have elapsed since the Cambrian period, but this,

judging from the small amount of organic change since the commencement of the Glacial epoch, appears a very short time for the many and great mutations of life ...<u>the world at a very early period was subjected to more rapid and violent changes in its physical conditions than those now occurring; and such changes would have tended to induce changes at a corresponding rate in the organisms which then existed</u>.[37]

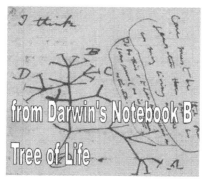

a.f. theguardian.com

Evolution by environment?!   What happened to slow and gradual natural selection?

Molecular Evolution?

It is often argued that molecular similarities among various animals, indicates frogs-to-Frank evolution.   The following represents the percent similarity of relaxin, which widens the birth canal:

Pig v. Human 46%
Pig v. Rat   54%
Pig v. Shark   50-52%
Pig v. Skate   31%

Does this represent an evolutionary sequence? With respect to insulin, Christian Schwabe (*The Genomic Potential Hypothesis*) observes that, "... the variation between pig and carp or pig and shark is less than the difference between the pig and hystricomorph rodents [includes porcupines, guinea pigs]"[38] Schwabe, writing in *Trends in Biochemical Sciences*, makes this remarkable statement: "… we need not look for a common ancestor of mollusks and man because <u>one might never have existed</u>."[39]

landesbioscience.com

Evolution by Guessing

Darwin's work was largely based on surmise and conjecture, but what about up-to-date books on evolution? The next time you're in Barnes and Noble or Books-A-Million and pick up a book on evolution look for the following words: assume, assumption, suppose, scenario, if, presume, take for granted, supposition, conjecture, surmise, guess, speculate, presupposition and the like. Let's take John Tyler Bonner's 2013 work, *Randomness in Evolution* (Princeton University Press) as an example:

> We <u>assume</u> that from an evolutionary point of view these events did not all arise at the same time …[40]

> If there is, as we <u>assume</u>, an ever present selection for larger forms …[41]

> One would <u>assume</u> that natural selection is responsible for the general shape of the fruiting bodies …[42]

The complicated event that gave rise to the eukaryotic cell, we <u>presume,</u> involved the association, or fusion, of prokaryotic elements.[43]

…all we can do is <u>guess in very general terms,</u> and it is likely that mammals' social origin is more recent than that of insects.[44]

…why have not the ancestral, and <u>presumably</u> less efficient, species gone extinct? <u>We can only guess at the answer</u> to this question …[45]

Does Books-A-Million accept OEF's (Old Earth Fallacies) and millions of years? Is evolution a lost art or can Darwin's theory be improved? Does evolution have skills? - maybe not. All nations want to be freed — escape the Darwin Creed. The only place we can see evolution occurring is on a game like Spore:

How does Darwinism relate to catastrophism and Young Earth Science (YES)? We deal with these issues in the next chapter.

Notes:
1) "The Origin of the Species" by Shaun Hellend
http://www.poemhunter.com/poem/the-origin-of-the-species/
2) "2014 : What Scientific Idea Is Ready For Retirement? - Essentialism"
by Richard Dawkins,
http://www.edge.org/response-detail/25366
3) "Essentialism in Aristotle's biology" by Christopher Gill, *Critical Quarterly,* Dec. 2011, Vol. 53 Issue 4, pp. 12-20, p. 14.
4) "Evolutionary Essentialism" by Denis Walsh, *British Journal of the Philosophy of Science*, 57 (2006), 425–448, p. 425.
5) Ibid., p. 429.
6) Ibid., p. 431.
7) Quoted in Walsh, p. 436, emphasis added.
8) *The Flamingo's Smile: Reflections in Natural History* by Stephen Jay Gould, emphasis added.
http://books.google.com/books?id=whOUFTkj4d4C
9) *Organismic Evolution* by Verne Grant (W.H. Freeman, San Francisco, 1977), p. 22.
10) *The Neck of the Giraffe* by Francis Hitching (Ticknor & Fields, New Haven, CT, 1982), p. 61.
11) *The Age of the World* by J. Logan Lobley (Robert Ashley, London, 1914), p. 113, emphasis added.
12) *Darwin and the Darwinian Revolution* by Gertrude Himmelfarb (Chatto and Windus, London, 1959), p. 287, emphasis added.
13) "The Origin of Darwinism" by C. Darlington, *Scientific American*, Vol. 201, May 1959, p. 66, emphasis added.
14) "The Fall and Rise of Catastrophism" by Trevor Palmer (1996), emphasis added,

http://archive.is/7LuiT

15) *Catastrophism* by Richard Huggett (Verso, London, 1997), p. 96.

16) *Wonderful Life* by Stephen Jay Gould (W.W. Norton & Co., NYC, 1989), p. 14.

17) "Platypus proves even odder than scientists thought" by Ian Sample, http://www.theguardian.com/science/2008/may/08/genetics.wildlife

18) *Unknown Earth* by William Corliss (The Source Project, Glen Arm, MD, 1980), p. 629, emphasis added.

19) *The Science of Evolution* by William Stansfield (Macmillan, NYC, 1977), p. 571.

20) Ibid., p. 577.

21) Ibid., pp. 574-577.

22) "Darwin Retried" http://www.amazon.com/Darwin-Retried-Norman-Macbeth/dp/0855112816

23) *Evolution of Living Organisms* by Pierre-Paul Grasse (Academic Press, NYC, 1977), p. 8, emphasis added.

24) Ibid., p. 13

25) Ibid., p. 228.

26) Ibid., p. 243.

27) *Catastrophism* by Richard Huggett (Verso, London, 1997), pp. 104, 105.

28) Ibid., p. 154

29) Ibid., p. 198, emphasis added.

30) "Richard Dawkins & Daniel Dennett vs. Francis Collins & Benjamin Carson" http://www.youtube.com/watch?v=JPxGnN7RV1Y

31) Personal interview on Oct. 7, 1981.

32) *The Origin of Species* - 6th Edition by Charles Darwin Chapter 15 - Recapitulation And Conclusion, emphasis added. http://www.literature.org/authors/darwin-charles/the-origin-of-species-6th-edition/chapter-15.html

33) *Wonderful Life* by Stephen Jay Gould (W.W. Norton & Co., NYC, 1989), p. 57.

34) Ibid., p. 99

35) *The Neck of the Giraffe* by Francis Hitching (Ticknor & Fields, New Haven, CT, 1982), p. 19.

36) *The Origin of Species* by Charles Darwin (6th ed.), Ch. 6, last paragraph, emphasis added. http://archive.org/stream/theoriginofspeci02009gut/pg2009.txt

37) *The Origin of Species* - 6th Edition by Charles Darwin Chapter 10 - On The Imperfection Of The Geological Record, emphasis added. http://www.gutenberg.org/files/2009/2009-h/2009-h.htm

38) "Talking Point: On the validity of molecular evolution" by Christian Schwabe, *Trends in Biochemical Sciences*, July 1986, pp. 280-283, p. 281.

39) Ibid., p. 282, emphasis added.

40) *Randomness in Evolution* by John Tyler Bonner (Princeton University Press, Princeton, NJ, 2013), p. 75, emphasis added.

41) Ibid., p. 19, emphasis added.

42) Ibid., p. 59, emphasis added.

43) Ibid., p. 21, emphasis added.

44) Ibid., p. 36, emphasis added.

45) Ibid., p. 59, emphasis added.

Chapter 4
The Day the Earth Died

A waste of waters ruthlessly
Sways and uplifts its weedy mane
Where brooding day stares down upon the sea
In dull disdain
- James Joyce

The floods, by Nature enemies to land,
And proudly swelling with their new command,
Remove the living stones, that stopped their way,
And gushing from their source, augment the sea …
The solid piles, too strongly built to fall,
High o'er their heads, behold a watery wall:
Now seas and Earth were in confusion lost;
A world of waters, and without a coast.
- Ovid (*Metamorphoses*)[1]

The Renaissance of Catastrophism

In 1923 J. Harlen Bretz advanced the concept that a great flood of water stormed through Montana, Idaho, Oregon and Washington into the Cascade Mountains and finally to the Pacific ocean. This event created the Channeled Scablands. The flood cut deep gorges (coulees) in solid basalt - Grand Coulee has walls up to 900 feet high. Forty years later, in 1963, the International Association for Quaternary Research met in the U.S. and sent a telegram to Bretz ending "We are now all catastrophists."[2] Bretz received the prized Penrose Medal (highest geology award) and claimed, "Perhaps, I can be credited with reviving and demystifying legendary Catastrophism and challenging a too rigorous Uniformitarianism."[3]

a.f. usgs.gov

NOVA created a powerful program on Bretz and the Channeled Scablands, *Mystery of the Megaflood*. Bretz' ideas were initially dismissed by his peers, but he stood his ground and defended catastrophism. Could this same procedure (glacial lake breach) have made Grand Canyon in like manner?

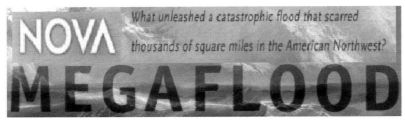

a.f. pbs.org

Suppose a giant megaflood encompassed the whole world. A planet-wide convulsion would imply that most rock strata were formed suddenly. A compressed geological time scale supports Young Earth Science (YES). According to Richard Huggett, Senior Geography Lecturer at the University of Manchester, "Before 1980, few geoscientists were willing to accept the idea that catastrophes have been of overriding importance in determining the history of the Earth."[4]

René Thom

a.f. wikipedia

Norman MacBeth, Johan Kloosterman and Otto Schindewolf among others wrote in *Catastrophist Geology*. Mathematician René Thom, who won the Fields Medal in 1958, wrote "Plate Tectonics and Catastrophe Theory" in this journal.[5]

Traditionally, the terms *uniformitarian* and *catastrophist*, have been used as antonyms. In 1832 philosopher of science William Whewell said,

> Have the changes which lead us from one geological state to another been, on a long average, uniform in their intensity, or have they consisted of epochs of paroxysmal and catastrophic action, interposed between periods of comparative tranquility? These two opinions will probably for some time divide the geological world into two sects, which may perhaps be designated as the *Uniformitarians* and the *Catastrophists*.[6]

A catastrophe is quick violent action on a massive scale. Michael Wysession (Geology Professor, Washington University, St. Louis) rightly stated,

> ... with the discovery of the impact that killed off the dinosaurs 65 million years ago, history has taken a slightly different view of catastrophism, and the works of people like Cuvier. For awhile, they were scorned and somewhat denigrated ...Yet it's the big volcanoes, the Krakatoas, or the Yellowstone super volcanoes, that

do tremendous damage and change, over very rapid timescales.[7,8]

Stephen Jay Gould showed proper respect for alternative science:

> If we equate uniformity with truth and relegate the <u>empirical claims of catastrophism to the hush-hush unthinkable</u>... then we enshrine one narrow version of geological process as true *a priori*, and <u>we lose the possibility of weighing reasonable alternatives</u>.[9]

a.f. channer.tv

Marilyn vos Savant (born in St. Louis) has been the *Guinness Book of World Records* record holder for the highest IQ. She said that, "If the Earth were smooth, all the land would be covered with seawater to a depth of about 8810 feet ..."[10] That's what a total flood on Earth would look like, but what about Mars? There are massive reservoirs of ice beneath the surface of Mars. If the ice melted it would create a planetary flood. BBC science editor David Whitehouse explains the Great Mars Flood:

> All over its surface there is evidence that in the distant past copious amounts of water flowed. We can see dried up river lakes, ancient shorelines, and vast, empty canyons. ... If Mars were to become much warmer for

some reason and the ice melted, <u>it would drench the planet to an average depth of between half and one kilometer.</u>[11]

Could our planet have also experienced such planet-wide destruction?

Geologists and Catastrophism

Some claim that a global calamity with massive impact requires a complete revision of geological science. In contrast, William Stokes and Sheldon Judson state in their Geology text *Introduction to Geology* (Prentice-Hall) that,

> A catastrophist might contend that the twisting and breaking of strata, the transportation of huge blocks of rock, the violent cutting of canyons, and the wholesale destruction of life is within the power of a great universal flood - and he would be right.[12]

George Gaylord Simpson made a similar statement:

> … <u>there is no a priori or philosophical reason for ruling out a series of natural worldwide catastrophes as dominating earth history.</u> However, this assumption is simply in such flat disagreement with everything we now know of geological history as to be completely incredible.[13]

During the current Renaissance of Catastrophism, should we not change this evaluation from "incredible" to "conceivable"?

Austrian geologists Alexander Tollmann and Edith Kristan-Tollmann proposed in 1994 that a large comet split up into seven pieces and hit the Earth. This caused megatsunamis around 7640 BC causing continental wide flooding. So, the Tollmann's hold to a nearly global flood.[14] Georges Cuvier (d. 1832), the Father of Paleontology, held to a global catastrophe that occurred a few thousand years before Christ:

... this revolution had buried all the countries which were before inhabited by men and by the other animals that are now best known; that the same revolution had laid dry the bed of the last ocean, which now forms all the countries at present inhabited ...[15]

Alcide d'Orbigny (d. 1857) was Professor of Palaeontology at the Paris Muséum National d'Histoire Naturelle.[16] d'Orbigny not only rejected evolution, but held to multiple cataclysms on the Earth that destroyed all life.[17]

In the late 1800's physicists were pushing for a younger earth (as young as 10M or 20M) and the geologists were hoping for an older earth (~100M). John Perry said in 1895 in *Scientific American*,

> Some physicists tell them that the flaw in the geologists' reasoning consists in their not taking into account the <u>much greater tidal actions of the past.</u> <u>When tides rose and fell many hundreds of feet, and</u> <u>swept over tens or hundreds of miles of foreshore, there</u> <u>must undoubtedly have been a more rapid formation of</u> <u>sedimentary rock</u> ... We acknowledge that all nature's actions were on the whole, <u>possibly more intense in the</u> <u>past.</u> We know from [George Darwin's research] that the moon was undoubtedly nearer the earth in Paleozoic times and the <u>tide influence was therefore greater.</u>[18]

If megafloods happened hundreds of miles from the coasts we have near continental flooding! As a consequence, catastrophic geologic features would result. Such formations formed *quickly*. Thus, catastrophism and a young earth are intimately linked.

Geologist Jon Erickson contends that a Global Catastrophe **is** possible:

If the [large] asteroid landed in the ocean, upon impact it would produce a conical-shaped curtain of water. Billions of tons of seawater would be splashed high into the atmosphere. The meteorite would instantly evaporate massive quantities of seawater, saturating the atmosphere with billowing clouds of steam. Thick cloud banks would shroud the planet, cutting off the Sun and turning day into night. The most massive tsunamis ever imagined would race outward from the impact site. The waves would completely traverse the world. When striking seashores, they would travel hundreds of miles inland, devastating everything in their paths, making such impacts among the most devastating, if rare, geologic hazards on Earth.[19]

a.f. meteorite.org

There are hundreds of impact structures on the earth. Many of these may have had a significant geologic effect. The Manicouagan impact structure in Quebec, Canada is 60 miles wide![20] If we include the "confirmed," "most probable" and "probable" categories, there are at least 293 impact structures on earth.[21]

Will Catastrophism Survive?

Comet impacts may not only be the cause of vast destruction on the earth, but may also provide a source of water for a global catastrophe. In 2000 comet LINEAR(C/1999 S4) broke apart as it passed near the Sun. Comet LINEAR was likely made up of water with the same isotopic composition as Earth water (H_2O, not HDO). This supports the thesis that comet impacts helped provide water for global catastrophe just a few thousand years ago.[22] Many geologists admit to continent wide flooding that occurred hundreds of millions of years ago. Some accept a global sea-level rise of 225m above present sea level, while others propose a rise as much as 600m![23] A continent flood to a global ocean – is that such a giant leap? Could a global flood have killed the Earth just thousands of years ago, instead of millions?

Often the possibility of catastrophic events is downplayed when considering formations that have long been thought to have been produced at a snail's pace. In the early 1900's a whale skull (*Megapteramiocaena*) was found in diatomite beds in Lompoc, California. Remington Kellogg explains, "The beds of diatomaceous earth at Lompoc furnish a very extensive marine fauna, composed mainly of fishes and to a less extent of birds, cetaceans, and pinnipeds."[24] It had been supposed by some geologists that diatomaceous earth formed very slowly. The tale of this whale and other fossils, including birds, is that this formation came about via a rare episodic event. Catastrophism and Young Earth Science (YES) go hand-in-hand.

Myths are Legendary

Kevin Krajick chronicles the myth and geology trend:

The movement traces in part to the 1980s, when scientists realized that the slow march of geologic time is sometimes punctuated by [massive] catastrophes, such as the giant meteorite that wiped out dinosaurs 65 million years ago. After this was accepted, some (usually those with tenure) felt freer to wonder if <u>near-universal myths of great floods and fires implied that such disasters also have punctuated human time</u>. ... People on the volcanic island of Kadavu, Fiji, have a suggestive legend about a big mountain that popped up one night, and locals say they have heard rumbling from the main cone recently. In 1998, [Patrick] Nunn [geoscientist at the University of the South Pacific in Suva, Fiji] and others investigated the volcano but <u>decided on preliminary evidence that it had not erupted for 50,000 years. The island has been inhabited for only 3000 years</u>, so they concluded that the myth was imported. Months later, a new road cut revealed pot shards under a meter-deep layer of ash. <u>"The myth was right, and we were wrong,"</u> says Nunn.[25]

This also, gives us a hint that our dating methods may be seriously flawed. Traditional geology gave a date of 50K years, yet this episode occurred within the last 3K years!! There is even a book *Myth and Geology*:

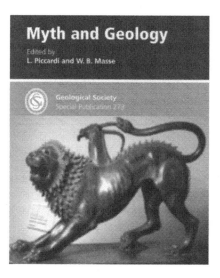

The Aztecs, Olmecs and Maya held that the interior of the Earth is filled with water. Could this represent a dim memory that this planet was killed by a watery inundation just thousands of years ago?[26] Native American tradition supports the view that the Grand Canyon was formed by a Megaflood, similar to the one that occurred in the Pacific Northwest. Both the Navajo and the Hualapai have legends that the Grand Canyon was formed by a huge flood.[27, 28]

Old King Coal

Did coal form at a snail's pace or rapidly? Robert Gastaldo and Carleton Degges state,

> A localized sandstone split in the Mary Lee coal [Pennsylvanian of Alabama] ... has a channel-form geometry and preserves a concentrated log-and-gravel (pebbles, cobbles, and boulders) assemblage ...Gravel lithotypes within and amongst rooting structures of lycopsid, cordaitean, and calamitean trees are indicative of an Appalachian orogenic provenance, and support an allochthonous [transported] origin for some of the logs. ... and are interpreted to represent an ancient log jam.[29]

The logs show a preferred orientation to the current direction which supports this claim. Traditionally, many thought that coal is formed in peat swamps. Many geologists hold that coal was formed by transported vegetation. That is, the coal was made in an allochthonous manner. Catastrophes can produce log jams in a flash.

Karl Alfred von Zittel observes,

> In France ... the theory of sedimentation [allochthonous coal formation] is strongly supported. Grand-Eury, the author of an excellent work published in 1882, upon the flora of the <u>Carboniferous formation of Central France, came to the conclusion that the coal-seams had originated by deposition in lake depressions surrounded</u>

by woods. Five years later, the *Etudes* of Henry Fayolon the Coal-deposits of Commentry brought forward a strong chain of evidence in favor of sedimentation from water. <u>Fayol shows how the pebbles, sand, mud, and plant detritus borne in suspension by rivers subside according to their weight, and arrange themselves as independent layers of sediment</u>. The coarser pebbles are deposited near the shore, usually with a distinct slope, while the light plant detritus is carried far out and deposited almost horizontally. ... It is in no small measure due to the prestige of this gifted geologist that <u>the sedimentation theory is held by the majority of French geologists at the present day [1901]</u>.[30]

Writing in *The geology of Svalbard* (North of Norway), Walter Harland et al state,

The Nordkapp Formation, 230m [thick] ... can be split into an upper and a lower unit. The upper unit contains interbedded conglomerate, sandstone and coaly shales. At least three coal seams are also present ... Coal seams, 10-60 cm thick, occur 15-40m below the top, in a sequence of carbonaceous black shales with plant remains and rare pyrite nodules. ... The transported plant remains and the absence of true seat-earths suggest that the <u>coals are allochthonous</u>.[31]

Hu Yicheng Liao Yuzhi (China University of Geosciences) and Li Zhaoming (Geological and Mineral Bureau of Henan Province) conclude regarding the Yiluo Coalfield in Henan Province that,

Late Carboniferous <u>allochthonous coal</u> is characterized by silty grains and shows similar <u>turbidity sequence</u> [rapid gravity flow sediments] ... The bottom strata of allochthonous coal, <u>not including root earth</u>, comprise coarse grained quartz sandstones with <u>Brachiopoda fossils and trace fossils</u>, indicating marine environments. ... <u>allochthonous coal</u> is found to be deposited when a large number of Carboniferous materials were <u>transported</u> to the areas below the medium tidal level or more by <u>storm events</u>.[32]

In other words, the coal was formed catastrophically. Turbidites form quickly! In Perry County in Ohio a quartzite boulder was found in a coal seem that weighted over 400 pounds!! This fact surely favors the catastrophic origin of coal.[33]

Quick Islands, Mountains and Canyons

Amazingly, Surtsey Island appeared off the coast of Iceland in 1963 in just a matter of months. Surtsey reached its maximum size of 2.7 km² in 1967. The island has a number of birds and plants. Surtsey has insects, spiders and even earthworms, probably brought there by birds. It does not take millions of years to form a new land.[34]

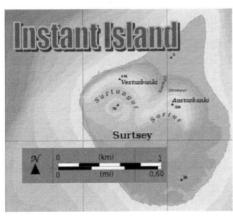

a.f. wikipedia

Does it take long for a mountain to form? In 1943, Paricutin Volcano started erupting (200 miles west of Mexico City). In just ten weeks it reached a height of 1,100 feet![35] Within one year the lava covered 6,200 acres.[36]

In 2002, Canyon Dam (Comal County, Texas) had to flood the Guadalupe River Valley in order to avoid catastrophic failure of the dam. The flood picked up meter-wide boulders. It also created a 2.2 km long and 7m deep canyon in the bedrock in just 3 days! Michael Lamb (Asst. Prof. Geology, Caltech) comments, "We know that some big canyons have been cut by large catastrophic flood events during Earth's history."[37]

Thick Deposits Form Fast

Up to 200 meters (656 ft.) of pyroclastic flow deposit accumulated in only 3.5 hours when Mount Pinatubo (Philippines) erupted in 1991.[38] Geologic deposits can form very quickly!

a.f. geolsoc.org.uk

The Deccan Traps in India consists of 10K km^3 of volcanic deposits. What triggered this activity? A number of works suggest that impacts of extraterrestrial objects may have caused volcanic eruptions:

The Cosmic Winter by Victor Clube and Bill Napier

The Mass Extinction Debates, edited by William Glen

Catastrophism, Neocatastrophism and Evolution by Trevor Palmer.[39]

a.f. foxnews.com

In 2013 a "mud volcano" rose from the seafloor near Gwadar, Pakistan. A new island was created quite suddenly. It is about 60 to 70 feet high. The surface is a mixture of mud, fine sand, and solid rock.[40]

The 1980 eruption of Mount St. Helens created significant geologic structures. The landslide and debris avalanche covered 23 square miles to a depth of up to 600 feet!! Lahars filled the Columbia River to a depth of 26 feet. Pyroclastic flows covered six square miles to a maximum depth of 120 feet![41] The volcano Monowai, near Tonga, experienced huge changes in height in just two weeks in 2012. Lava flows raised one section by 79m.[42]

The Upside-Down Birthday Cake

Native American scholar Vine Deloria Jr. rightly observes,

> The rocks are not always found in the proper order; sometimes a formation has simpler fossils on the top and more complex ones on the bottom. To solve this problem, geologists invented the "overthrust." This term is applied to formations in which, in theory, some layers of older strata are pushed up over younger strata, disrupting the sequence of fossils. This concept has been liberally applied when the fossils do not line up in a coherent evolutionary sequence, even when there has been no evidence that strata have been disturbed.

Overthrust is thus often used to hide anomalies and explain inconsistencies in the progression of fossils.[43]

These overthrusts may be real, but when they involve regional masses of rock, the only reasonable cause is some cataclysmic event - thus destroying the uniformitarian empire.

Summarizing a piece in the *Geological Society of America Bulletin* by Paul Washington and Raymond Price on the paradox of large overthrusts, Science-Frontiers state,

> In many parts of the world, older rocks are found on top of younger rocks. Obviously the Principle of Superposition is contradicted in such situations. Two possible explanations exist for these inverted strata: (1) The dating of the rocks is incorrect; or (2) Geological forces somehow slid the older rocks over the top of the younger rocks. ... Three potential solutions have been proposed: (1) Lubricate the sliding surfaces with water under high pressure (the pore-pressure approach); (2) Allow the thrust block to slide, not as a unit, but in small discrete areas at different times (the dislocation approach); and (3) Push the thrust block not only from the rear edge but along the top surface (the tapered wedge approach). In fact, all three solutions may apply; but there is no consensus so far. Each solution has problems.[44]

According to William Pierce,

> The Heart mountain fault [Wyoming] movement was an extreme rapid, cataclysmic event that created a large volume of carbonate fault breccia ... The calcibreccia dikes ... were injected at essentially the same time [as the fault movement] and show a common mechanism, namely, lithostatic pressure due to burial by a rapidly accumulating cover of volcanic rocks ... The dikes confirm an earlier conclusion ... that the Heart Mountain fault movement was a cataclysmic event ...[45]

If the Heart Mountain overthrust formed rapidly, could it be that other "wrong order" strata were formed in a catastrophic manner?

Grand Mystery – Grand Canyon

Wikipedia describes an interesting idea on the origin of the Grand Canyon:

> ... an alternative theory is that roughly 5.4 million years ago, the Colorado's flow again shifted into Hopi Lake. Hopi Lake overflowed across the Colorado Plateau in a massive flood, eroding the Grand Canyon in a much shorter time period than previously believed.[46]

a.f. YouTube

In a publication of Grand Canyon Association, Norman Meek and John Douglass propose that,

> ...the Colorado River could have developed by the episodic downstream extension of its trunk channel from multiple lake-overflow events between ~10 and 4 Ma. ... we revisit a hypothesis than was originally suggested for the entire Colorado River drainage system by Eliot Blackwelder in 1934: lake overflow. ...we propose that a large lake spilled westward across

the Kaibab Plateau initiating Grand Canyon incision. The deep-water lake that would have spilled over the Kaibab Plateau could be called lake Bidahochi ...[47]

a.f. maricopa.edu

Eliot Blackwelder (d. 1969) was the chair of the Geology Department at Stanford University and presided over the Geological Society of America.[48]   Could the epic destruction that made the Grand Canyon have happened just thousands of years ago?

Establishment geologists now admit that huge lava dams in the Grand Canyon created temporal lakes that burst and made existing canyons deeper. One such massive flash flood carried 37 times more water than the largest ever recorded on the Mississippi. Geologists now hold that some parts of Grand Canyon maybe no more than 100K years old, much younger than prior estimates of millions of years. Some of these lava dams were over 1500 feet tall![49]

Sanjeev Gupta and Jenny Collier (Imperial College London) propose that a megaflood carved the English Channel.[50] They calculated that 1 million cubic meters of water per second catastrophically flowed as recently as 200K years ago. This cataclysm carved valleys up to 10 km wide and 50 m deep, with a flow rate a hundred times larger than the Mississippi. The source of the water was a glacial lake. In 1985, marine geologist Alec Smith of Bedford College in London came to the same conclusion! Could the rejection of Smith's theory have been the result of the anti-catastrophist bias in the scientific community? Could a similar megaflood have carved out the Grand Canyon?

Tree Rings are the Key

If an Episode of Global Calamity did occur, when did it happen? According to Edmund Schulman (Div. of Geological Sciences, Calif. Inst. of Tech.),

> ... *Sequoia gigantea* trees ... may enjoy perpetual life in the absence of gross destruction, since they appear immune to pest attack. ... Does this mean that <u>shortly preceding 3275 yr ago (or 4000 yr ago,</u> if John Muir's [figure is] correct) <u>all the then living giant sequoias were wiped out by some catastrophe</u>?[51]

Intelligent Catastrophsim

We are introducing Intelligent Catastrophism (IC) - the idea that rapid geologic action implies a young world. If much of the geologic record formed quickly and there are no large time gaps between strata (see the next chapter), then the whole of geologic time is much shorter than standard science claims. It is fascinating to note that Gould used "intelligent catastrophism" in a slightly different context.[52] How does Intelligent Catastrophism brighten our understanding of the history of life we find in the fossils? We tackle this issue in the next chapter.

Notes:

1) *Metamorphoses* by Ovid, translated by Sir Samuel Garth, John Dryden, et al

http://classics.mit.edu/Ovid/metam.mb.txt

2) *The New Catastrophism* by Derek Ager (Cambridge University Press, 1993), p. 19, emphasis added.

3) "GSA Medals and Awards," Geological Society of America, *GSA News & Information*, Vol. 2, 1980, p.40.

4) *Catastrophism* by Richard Huggett (Verso, London, 1997), p. 113.

5) "Catastrophist Geology"

http://www.catastrophism.com/cdrom/pubs/journals/catgeo/index.htm

6) quoted in *Catastrophism* by Richard Huggett (Verso, London, 1997), p. xviii, emphasis in original.

7) "How the Earth Works – Course Introduction"

http://teachingcompany.12.forumer.com/a/course-introduction_post1128.html

8) "How the Earth Works - Geologic History - Dating the Earth"

http://teachingcompany.12.forumer.com/a/2-geologic-history-dating-the-earth_post1377.html

9) *Time's Arrow, Time's Cycle* by Stephen Jay Gould (Harvard Univ. Press, Cambridge, MA, 1987), p. 114, italics original, emphasis added.

10) "Ask Marilyn" by Marilyn vos Savant, *Parade*, July 9, 2000, p. 14.

11) "Mars ice could flood planet" by David Whitehouse

http://news.bbc.co.uk/2/hi/science/nature/2013114.stm

Accessed 12/26/02, emphasis added.

12) *Introduction to Geology* by William Lee Stokes and Sheldon Judson (Prentice-Hall, Englewood Cliffs, New Jersey, 1968), p.296.

13) "Historical science" by George Gaylord Simpson in *The Fabric of Geology* ed. by Claude Albritton (Addison-Wesley, Reading, MA, 1963), p. 32, emphasis added.

https://archive.org/details/fabricofgeology007224mbp

14) "Tollmann's hypothetical bolide"

http://en.wikipedia.org/wiki/Tollmann%27s_hypothetical_bolide

15) Quoted in *Catastrophism* by Richard Huggett (Verso, London, 1997), p. 94.

16) "Alcide d'Orbigny"

http://en.wikipedia.org/wiki/D%27Orbigny

17) *Catastrophism* by Richard Huggett (Verso, London, 1997), p. 95.

18) "The Age of the Earth" by John Perry (1895) in *Determining the Age of the Earth* (Scientific American, 2013), p. 78, emphasis added.

http://dafix.uark.edu/~danielk/Darwin/AOTE_issue.pdf

19) *Quakes, Eruptions and Other Geologic Cataclysms* (*Revealing the Earth's Hazards*, Rev. Ed.) by Jon Erickson (Checkmark Books, NYC, 2001), p. 245, 246.

20) Ibid., p. 224

21) "Impact Database – History and Current Status"
http://impacts.rajmon.cz/IDhistory.html

22) "A Taste for Comet Water"
http://science1.nasa.gov/science-news/science-at-nasa/2001/ast18may_1/

23) "Reconstructing Eustatic and Epeirogenic Trends from Paleozoic Continental Flooding Records" by Thomas Algeo and Kirill Seslavinsky *Sequence Stratigraphy and Depositional Response to Eustatic, Tectonic and Climatic Forcing, Coastal Systems and Continental Margins* ed. by Bilal Haq, Volume 1, 1995, pp 209-246, p. 209.
http://link.springer.com/chapter/10.1007/978-94-015-8583-5_8

24) "Description of the skull of Megapteramiocaena, a fossil humpback whale from the Miocene diatomaceous earth of Lompoc, California" by Remington Kellogg, *Proceedings of the United States National Museum*, 61(2435): 1-18, 1922
http://si-pddr.si.edu/dspace/handle/10088/15207

25) "Tracking Myth to Geological Reality" by Kevin Krajick, emphasis added.
http://www.freerepublic.com/focus/f-news/1516400/posts

26) *Lost Discoveries* by Dick Teresi (Simon & Schuster, NYC, 2002), pp. 264, 265.

27) *Doomsday: the Science of Catastrophe* by Fred Warshofsky (Reader's Digest Press, NYC, 1977), p. 81.

28) "Hualapai Indian Legend"
http://grand-canyon-vacation-information.com/hualapai-indian-legend.html

29) "Sedimentology and paleontology of a Carboniferous log jam" by Robert Gastaldo and Carleton Degges, *International Journal of Coal Geology* 69 (2007): 103–118, p. 103.

30) *History of Geology and Paleontology to the End of the Nineteenth Century* by Karl Alfred von Zittel (Walter Scott, London, 1901), trans. by Maria Ogilvie-Gordon, p. 242, emphasis added.

31) *The geology of Svalbard* by Walter Harland, Lester Anderson, Daoud Manasrah and Nicholas Butterfield (The Geological Society, Bath, UK, 1997),pp. 216, 217, emphasis added.

32) "Late Carboniferous Allochthonous Coal of Yiluo Coalfield in Henan Province" by Hu Yicheng Liao Yuzhi and Li Zhaoming, emphasis added.
http://en.cnki.com.cn/Article_en/CJFDTOTAL-DQKX806.008.htm

33) "On the Occurrence of a Quartz Boulder in the Sharon Coal" by Edward Orton in *Unknown Earth* by William Corliss (The Source Project, Glen Arm, MD, 1980), pp. 152, 153.

34) "Surtsey"
http://en.wikipedia.org/wiki/Surtsey

35) *Quakes, Eruptions and Other Geologic Cataclysms* (*Revealing the Earth's Hazards*, Rev. Ed.) by Jon Erickson (Checkmark Books, NYC, 2001), p. 77.

36) "Parícutin" [Volcán de Parícutin]
http://en.wikipedia.org/wiki/Par%C3%ADcutin

37) "Geologist investigates canyon carved in just 3 days in Texas flood"
http://www.geologytimes.com/research/Geologist_investigates_canyon_carved_in_just_3_days_in_Texas_flood.asp

38) *Super-eruptions: global effects and future threats* by S. Sparks and S. Selfetal (Geological Society of London, 2005)
www.geolsoc.org.uk/supereruptions

39) "The Fall and Rise of Catastrophism" by Trevor Palmer (1996),
http://archive.is/7LuiT

40) "Earthquake Births New Island off Pakistan" by William Barnhart et al
http://earthobservatory.nasa.gov/NaturalHazards/view.php?id=82146

41) "1980 eruption of Mount St. Helens"
http://en.wikipedia.org/wiki/1980_eruption_of_Mount_St._Helens

42) "Rise and fall of underwater volcano revealed" by David Shukman
http://www.bbc.co.uk/news/science-environment-18040658

43) *Evolution, Creationism, And Other Modern Myths* by Vine Deloria Jr. (Fulcrum Pub., Golden, CO, 2002), p. 92

44) "The Mechanical Paradox In Thrust Faulting"
http://www.science-frontiers.com/sf074/sf074g11.htm

45) "Clastic Dikes of Heart Mountain Fault Breccia, Northwestern Wyoming, and Their Significance" by William Pierce, *Geological Survey Professional Paper* 1133, 1979, pp. 1-25, p. 1, emphasis added.

46) "Colorado River"
http://en.wikipedia.org/wiki/Colorado_River
Accessed 11/22/2011.

47) "Lake Overflow: An Alternative Hypothesis for Grand Canyon Incision and Development of the Colorado River" by Norman Meek and John Douglass in *Colorado River: Origin and Evolution* ed. by Richard Young and Earle Spamer, proceedings of a symposium held at Grand Canyon
National Park in June 2000 (Grand Canyon, Arizona: Grand Canyon Association, 2001), pp. 199-204, pp. 199, 202.
http://geomorphology.sese.asu.edu/Papers/31-lake_overflow-an_alternative_hypothesis.pdf

48) "Eliot Blackwelder (1880-1969) and Lake Overflow re-proposed 1934"
http://www2.pvc.maricopa.edu/~douglass/v_trips/grand_canyon/introduction_files/Part3.html

49) "Is the Grand Canyon a Geologic Infant?" by Ed Stiles
http://uanews.org/story/grand-canyon-geologic-infant

50) "Megaflood carved the English Channel" by Paul Marks, *New Scientist*, Vol. 195, Issue 2613, July 21, 2007, p. 11.
http://www.newscientist.com/article/mg19526134.200-megaflood-carved-the-english-channel.html

51) "Longevity under Adversity in Conifers" by Edmund Schulman, *Science* Mar. 26, 1954, Vol. 119, No. 3091, pp. 396-399, p. 399, emphasis added.
http://www.wmrs.edu/projects/WMRS%20history/history%20documents/science%201954.pdf

52) *Time's Arrow, Time's Cycle* by Stephen Jay Gould (Harvard Univ. Press, Cambridge, MA, 1987), p. 115.

Chapter 5
Who's SERGA?

Geology isn't a real science.
- Sheldon (*The Big Bang Theory*)

Old Earth, worn by the ages, wracked by rain and storm, exhausted yet ever ready to produce what life must have to go on!
- Charles de Gaulle (d. 1970)

Different Rocks or Different Life?

Does the Geologic Column really tell us the story of the transformation of life from simple to complex through time? Or, does it primarily give us the record of a significant environmental failure that destroyed the Earth just thousands of years ago? Consider the history of the geologic column. Adam Sedgwick's demarcation of the Cambrian system was based on the lithology (types of rock), not the fossils. The name came from Welsh tribes. The Devonian is based on the rocks of Devonshire in southern England. The Carboniferous (Mississippian and Pennsylvanian) was initially based on lithology - coal beds. The Permian was named after great limestone layers in Perm, Russia. Friedrich von Alberti named the Triassic after rocks in Germany with a characteristic three-fold division and not based on the fossils. The Jurassic is based on rocks of the Jura Mountains of northern Switzerland. The Cretaceous is based on chalk beds near Paris. Again, lithology and not fossils are the primary description.[1]

Whence Fossil Order?

Many claim that the fossil record shows an order in the life forms that precludes viewing most of the sediments as part of one catastrophic episode – a **S**ingular **E**poch of **R**apid

Geologic Activity (SERGA). Evolution does not explain the sequence of the fossils. The general order of the paleontological data can be explained by ecological zonation, early burial of marine animals (95% of all fossils are shallow marine organisms), hydrodynamic sorting (diameter, sphericity and density) and behavior and the higher mobility of vertebrates. Birds are found in the rocks dated more recently – they can fly away from many kinds of disasters. The Cambrian Explosion exhibits a great diversity of life and is consistent with a global catastrophe. Evolution should show a gradual increase in the number of phyla – the Cambrian Explosion contradicts this. Darwin expected many intermediate forms and yet they have never been discovered. Thus the order goes against evolution since the missing links are not even part of the fossil order! The order expected by evolution should start like this (younger to older):

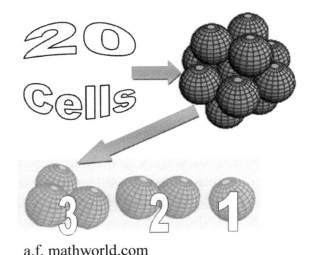

a.f. mathworld.com

Mark Patzkowsky and Steven Holland explain the meaning of fossil order:

> Stratigraphic paleobiology … is built on the premise that the <u>distribution of fossil taxa in time and space is controlled not only by processes of evolution, ecology, and environmental change, but also by the stratigraphic processes</u> that govern where and when sediment that might contain fossils is deposited and preserved.[2]

Let's take this list and remove evolution: ecology, environmental influence and stratigraphic processes. Evolution is superfluous.

In the journal *Lethaia*, Valentin Krassilov stated,

> Causal biostratigraphy means an approach to stratigraphic problems based on ecosystem analysis of interrelations between geological events and organic evolution. The succession of ecosystems is controlled mainly by climatic cycles. Stratigraphic units correspond to palececosystems. ... [there is] an inherent catastrophic explanation ... [rock units are] controlled by the same events which had been responsible for the stratum boundaries. This catastrophic explanation had been widely accepted in pre-Darwinian time, and its modernized version [has been] supported by several authors ...[3]

The fossil record can be explained as a pattern of palececosystems. If the rock units represent ecosystems of the past, why do we need quarks-to-cosmonauts evolution at all? Krassilov continues, "... continuous 'evolutionary' series derived from fossil record can in most cases be simulated by chronoclines - successions of a geographical cline populations imposed by the changes of some environmental gradients."[4]

The vertical arrangement of fossils does not necessarily teach us the story of evolution. As Robert Spicer explains,

> Both fluid and biological sorting of potential plant megafossils during transport and deposition processes produce a pattern that has previously been overlooked. ... Vertical differences in species composition need not necessarily reflect temporal changes in vegetation.[5]

The two-dimensional nature of plant materials (leaves, some seeds etc.), changes in density (e.g. waterlogging), fragmentation and bacterial decay are issues that models of depositional sorting must consider. Different types of life live in different zones and exhibit vertical stratification: abyssal, pelagic, neritic, euphotic, littoral etc. This arrangement could help explain the order of fossils within the SERGA framework.

Paleontological Pancake Party!

Geologist, oceanographer and archaeologist Tjeerd van Andel speaks of the incompleteness of the geologic record, "If erosion and other ravages of time are the cause of the missing record, <u>one should expect the incompleteness to increase with age. This, however, is not the case</u> …"[6]

a.f. news.stanford.edu

If most of the rock record is part of SERGA, this is not a difficulty. Following mainstream thinking, as rocks during the Ordovician period are formed some erosion will destroy the Cambrian layers below. Thus, the oldest rocks will have least representation in the rock record, but this is not what we observe as van Andel pointedly remarks. During a period between deposits a new landform is created and we have new valleys and river channels. These erode into previous formations. This results in a rather complex array of depositional regions:

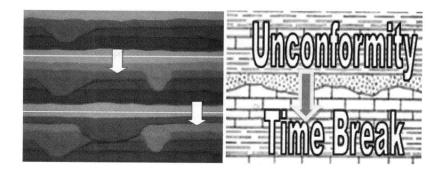

In contrast, if one layer was laid down soon after the prior formation, then we should see something like a stack of pancakes (e.g. as we see in the Grand Canyon). Suppose a limestone bed is soon overlain by a sandstone formation. In this case, there would be no time for valleys and erosional features to form.

Derek Ager admitted to the missing time in the Grand Canyon,

> Nowhere in the world is the record, or even part of it, anywhere near complete. Even in the Grand Canyon of the Colorado River and the adjacent sections along the Little Colorado River, surely the finest record of geological history anywhere on Earth, there are huge breaks. Notable is the complete absence of the Upper Carboniferous (Pennsylvanian) and of the Ordovician and Silurian Systems. Devonian strata are only present in local lenses.[7]

But what if the Grand Canyon represents continuous deposition? That would imply that "Ordovician time" never existed! The pancake like layers seems to show that these rock formations were part on one great catastrophe (SERGA), not multiple inundations.

Alfred de Grazia described problems with the Geologic Column,

> Rarely does one find even three of the ten geologic periods in their expected consecutive order. Moreover, 42% of Earth's land surface has 3 or less geologic

periods present at all; 66% has 5 or less of the 10 present; and only 14% has 8 or more geologic periods represented.[8]

What if this "missing time" never existed? Then the rock record may portray the activity of thousands and not millions of years. *Quantavolution - Challenges to Conventional Science* is a Festschrift in honor of de Grazia.

Edmund Spieker (Geology Dept., Ohio State Univ.) made this profound statement in 1956: "… how many geologists have pondered the fact that lying on the crystalline basement are found from place to place not merely Cambrian, but rocks of all ages?"[9] If most rocks were produced during a SERGA, this is exactly what we would expect!

In the Early Jurassic near Swansea (UK) Derek Ager tells about,

> … a conglomerate known as the "Sutton Stone." This has usually been interpreted as the basal conglomerate of a diachronous transgressive sea. It has been suggested, with very little fossil evidence, that <u>this conglomerate spans three to five ammonite zones and therefore up to five million years in time. I think it was deposited in a matter of hours or minutes.</u> …there are no bedding planes within the Sutton Stone in the coast sections. <u>Such matrix-supported conglomerates are characteristic of mass flow deposits</u>…[10]

It appears that the 5M years does not exist since Sutton Stone was formed catastrophically!

Rapid Geologic Activity

Many formations once thought to form over a long period of time are now explained catastrophically. In 1899 Frank Hall Knowlton noticed tropical plants in Yellowstone's "fossil forest." He found avocados, figs, acacias and breadfruits.

There were also trees of temperate climate (maples, dogwoods, elms etc.). What could cause such a strange mixture? Geologist William Fritz (Georgia State Univ.) looked at the mud flows of Mount St. Helens, which deposited trees upright, to help explain the Yellowstone phenomena. Thus, a great catastrophe mixed temperate and tropical trees. Prior theorists had calculated that the Yellowstone "fossil forest" represented over 20K years![11]

In Yellowstone, in the Lamar River Formation, there are several strata that have petrified wood. This has been interpreted in years gone by as a succession of fossil forests that took many thousands of years to form. However, as William Fritz explains,

> Volcaniclastic sedimentary rocks preserving the "fossil forests" of the Eocene Lamar River Formation in Yellowstone National Park were deposited by a complex alluvial system. Mud flows and braided streams ... transported plant parts, including some logs and stumps ... most of the logs are oriented in a particular direction, as in log jams ...[12]

So, these Yellowstone deposits were made catastrophically and quickly.

The Coconino Sandstone of the Grand Canyon is usually explained as wind deposited in a desert environment. Fossil tracks in the Coconino travel up, but never down. Could this be because these animals were escaping the vast devastation of a watery catastrophe?[13]

Paraconformities

In a paraconformity, strata are parallel with no sign of erosion. That is, the layers appear to be deposited continuously, but the fossils seem to tell another tale based on evolutionary PreSuppositions. Carl Dunbar and John Rodgers describe the Beargrass Quarry near Louisville, Kentucky where a "paraconformity between the Louisville limestone (Middle Silurian) and the Jeffersonville limestone (lower Middle Devonian). <u>This hiatus is represented by more than 3,000 feet of strata in parts of the Appalachian trough</u>."[14] This amounts to about 25M years of "missing time." Could it be that this "missing time" is just an illusion. That is, the Jeffersonville limestone was deposited immediately after the Louisville limestone – there is no time gap.

a.f. bbc.co.uk

Dunbar and Rodgers tell of another, "... paraconformity [that] separates Middle Silurian from the overlying Lower Devonian limestones in the Western Valley of Tennessee, where <u>the contact can be located only when fossils are found in the bounding layers</u>."[15] Dunbar and Rodgers continue in their discourse on paraconformities:

> In the Grand Canyon walls ... where the Redwall limestone can be dated as lower Mississippian and the underlying Muav limestone as Middle Cambrian, <u>we know that the paraconformity represents more than three geologic periods, yet the physical evidence for the break is less obvious</u> than [other boundaries] ... <u>Many large unconformities would never be suspected if it were not for such dating of the rocks</u> above and below.[16]

This represents a gap of about 150M years. The Ordovician, Silurian and Devonian are completely missing. What if they never existed? Then the Earth is young and Young Earth Science (YES) is true.

a.f. YouTube (Open Univ.)

This stack of coins is analogous to the rock record; or so we are told in a video segment from Open University. Suppose we have a Roman coin, a coin from WW II and a recent coin. How much time is missing between the coins? If we compare this to a stack of pancakes we know that one was laid down soon after the prior one. That is the whole stack formed rapidly. What do we actually find in the rock strata? Norman Newell said,

> The <u>Devonian rests paraconformably on Cambrian rocks over much of Montana</u> ... Many geologists would term this a disconformity, but <u>over large areas its recognition and evaluation depend solely on fossils.</u> Every experienced biostratigrapher can cite other examples of such paraconformities. In some cases many of these clearly represent long intervals of net erosion or non-deposition, as indicated by thinness of strata and gaps in the paleontological sequence. Judging from the thickness of more complete sequences found in adjacent regions <u>these stratigraphic breaks may be major unconformities equal to thousands of feet of sediment and millions of years of time.</u>[17]

Note the interpretation can go the other direction - the paraconformity demonstrates continuous deposition; therefore, the millions of years never existed.

Consider Signal Peak in Big Spring, Texas – are there millions of years between each layer of rock? Prentice-Hall published the "Foundations of Earth Science Series," one volume of which was *Geologic Time* by Don Eicher. Eicher admits that, "Some disconformity [paraconformity] surfaces … have no obvious relief, and appear very much like ordinary bedding planes. Without faunal evidence for a significant time break, some could – and undoubtedly do – go unnoticed."[18]   So we see that the rocks say no time gap, but Darwin's Army forces a large time period between the layers.

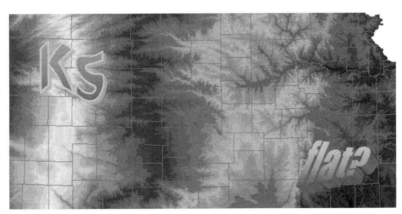

a.f. academic.emporia.edu

Is Kansas really as flat as a pancake? No, there are actually some channels and rises. We should find such erosion between strata if they are really millions of years apart in their origin. UK Geologist Myles Bowen (d. 2013), winner of the AAPG Pioneer Award, said regarding time signals in the sedimentary record, "Often the most difficult problem is to spot the gaps in apparently continuous sequences."[19]   What if the sequence really is continuous? Then, the whole geologic record may comprise a vastly shorter time frame than consensus science has imagined thus far.

Austrian geologist Eduard Suess (of Gondwana fame) said in 1906,

> …the regular and uniform character of the movements may be recognized from the concordant superposition of the more recent beds on those of much greater age. Of this there are numerous examples. Murchison, in describing the recent marine beds with Arctic shells at Ust-Waga on the Dwina, has pointed out their absolutely conformable superposition on the horizontal Permian sediments; he has also shown how at other places the latter sediments rest in perfect concordance on much older beds, so that the stratigraphical relations offer no hint of the great gap which occurs at the line of contact. That this should be the case may well be cause for astonishment, for some degree of erosion, weathering, or other alteration of the surface must have occurred in the interval …[20]

Could it be that the "missing time" never existed?

Alfred Selwyn relates the details of a paraconformity in the Canadian Rocky Mountains:

> East of the main divide, the Lower Carboniferous is overlain in places by beds of Lower Cretaceous age [that's a time gap of about 200M years!], and here again, although the two formations differ so widely in respect to age, one overlies the other without any perceptible break, and the separation of one from the other is rendered more difficult by the fact that the upper beds of the Carboniferous are lithologically almost precisely like those of the Cretaceous. Were it not for fossil evidence, one would naturally suppose that a single formation was being dealt with.[21]

Maybe it is ONE formation and the gap of millions of years is merely a fiction of Big Geo. Such paraconformities support the reality of a SERGA.

Multiple Thin Layers ➔ Old Earth?

It is often argued that repeated thin layers in the rocks implies lots of time. Lagerstätten are rapidly deposited sediments that generally have fine layers with exquisite fossil preservation. As Howard Feldman et al observe,

> Sites of <u>extraordinary fossil preservation</u> (Konservat Fossil- <u>Lagerstätten</u> of Seilacher, 1970) have long been exploited for their rich assemblages. Many sites yield diverse, abundant, and <u>excellently preserved fossils</u> that typically are not well preserved elsewhere. Nearly all of these Lagerstätten occur within <u>fine-grained, laminated rocks</u>. Depositional environments of these rocks are particularly difficult to interpret because few distinctive sedimentary structures are preserved other than <u>fine-scale laminations</u>.[22]

Lagerstätten Locations

Precambrian	Doushantuo, China	Ediacara, Australia
Cambrian	Burgess Shale, Canada	Chengjiang, China
	"Orsten," Baltic	Sirius Passet, Greenland
Ordovician	Soom Shale, South Africa	
Silurian	Ludlow Bonebed, Wales	
Devonian	Gilboa, USA	Gogo Formation, Australia
	Hunsrück Shale, Germany	Rhynie Chert, Scotland
Carboniferous	Mazon Creek, USA	
Jurassic	Karatau, Kazakhstan	LaVoulte-sur-Rhône, France
	Posidonienschiefer, Germany	Solnhofen, Germany
Cretaceous	Las Hoyas, Spain	Liaoning, China
	Santana, Brazil	Tlayúa, Mexico

data from palaeo.gly.bris.ac.uk

So, fine layers do not imply annual seasonal transitions. Furthermore, these fossils were deposited rapidly, otherwise they would have decayed. Another OEF (Old Earth Fallacy) is vanquished – that is, that many fine laminations imply a long time span.

In a letter to *Nature*, "Spontaneous stratification in granular mixtures," Hernán Makse et al state,

> Granular materials segregate according to grain size when exposed to periodic perturbations such as vibrations. ... [Also] when such a mixture is simply poured onto a pile, the large grains are more likely to be found near the base, while the small grains are more likely to be near the top. ... during [an] avalanche, the grains separate into a pair of static layers, with the small grains forming a sublayer underneath the layer of large grains.[23]

So, fine layers of sediment can form quickly.

In 1979, Kenneth Hsu (*Catastrophes, Dinosaurs and Evolution*) co-authored a paper in *Sedimentology* on "varves" [annual layer of sediment] in a Swiss lake (Walensee). These thin bands are often thought to represent annual layers. Hsu found five "varves" in a single year and concluded they formed rapidly by a catastrophic turbid underwater flow.[24]

Polystrate Fossils

A polystrate fossil extends through several strata. They occur in America, Canada, England, France, Germany, and Australia. Polystrate trees are often found near coal beds. According to Wikipedia, "A polystrate fossil is a fossil of a single organism (such as a tree trunk) that extends through more than one geological stratum." Such fossil trees in the "position of growth" (vertical) indicate rapid burial; otherwise, the tree would have rotted.[25]

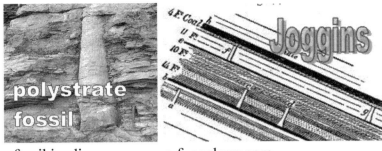

a.f. wikipedia   a.f. geology.com

The Joggins Formation (Pennsylvanian) of Nova Scotia is famous for its polystrate trees. Charles Lyell, in describing the Joggins formation, actually supports catastrophism: "… the erect trees consisting chiefly of large Sigillaria, occurring at ten distinct levels [there are actually 20], one above the other. …one trunk was about 25 feet high and four feet in diameter…"[26]

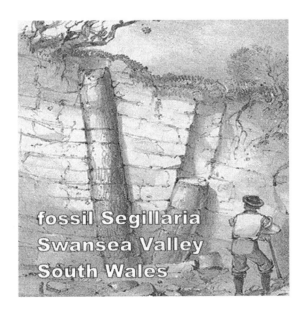

Derek Ager illuminates the implications of polystrate trees:

> Probably the most convincing proof of the local rapidity of terrestrial sedimentation is provided by the presence in the coal measures of trees still in position of life. Two Late Carboniferous trees stand in the garden of Swansea Museum [UK]…. Such standing trees are not uncommon in the Upper Carboniferous.[27]

We even find polystrate fossils in France:

Fig. 442. — Vertical trees in the Coal Measures sandstone, St.-Étienne, France. From Credner's *Elemente der Geologie.*

A Text-book of Geology – Historical Geology (1915)

Many hundreds of vertical cylindrical sandstone columns (sandstone pipes) are exposed in the Laguna area of New Mexico. They go through rocks of Jurassic age, including the Morrison formation. Some of these sandstone pipes are hundreds of feet long and up to 150 feet wide! Now does it not make more sense to a reasonable man that these sediments were all relatively soft when these pipes formed? Could these formations have remained soft for millions of years? If all these deposits were formed rapidly, then there is no enigma.[28]

Ager points out that escaping bivalves support catatrophism,

> Broadhurst, Simpson & Hardy (1980) also demonstrated episodic [catastrophic] sedimentation in the English Upper Carboniferous (Westphalian) by means of the non-marine bivalves, which are often common in these deposits. They plotted the number of "escape shafts" of these molluscs, when they had to burrow upwards to avoid fatal burial.[29]

These bivalves, striving to survive, had to dig upwards in order to get out of the thick deposit that had formed rapidly.

Turbidites

Turbidites are gravity flow formations often generated by earthquakes. Johns et al describe one such turbidite:

> Large-scale carbonate beds interbedded in basin-plain turbidites of the Eocene Hecho Group, northern Spain, are interpreted as giant turbidity-current deposits because of their internal organization and lateral extent. One such bed, the Roncal Unit, has been traced down the axis of the basin for 75 km and commonly reaches thicknesses in excess of 100 m.[30]

a.f. wikipedia

Thus, over 328 feet of sediment, including carbonates, were deposited in a very rapid catastrophic manner!

The Grand Banks earthquake of 1929 formed a turbidite bed covering over 100K square miles off Newfoundland![31] The Macigno Formation in Italy contains graded beds caused by turbidity currents. Some of these turbidites are over 20m thick.[32]

Can entire mountains move around catastrophically? Derek Ager says YES:

> … from the Apennines [Italy] came the idea of *cuneicomposti* or composite wedges of tectonic origin and submarine landslips formed under force of gravity … the mountain range formed as a series of successive

ridges, each consisting of a set of complex wedges of rock pushed upwards from the west. These were followed by <u>great submarine landslips ... In these, huge masses of rock up to the size of mountains, slid down into deeper water forming the great jumble of rocks</u> known as the *argillescagliose* or scaly clays. <u>Similar complex jumbles have been recognized in many other parts of the world such as Taiwan, Timor, Morocco and Turkey ...</u>[33]

Signs of SERGA

According to Ager deep-ocean vents are rare in the rock record,

> One interesting feature of modern oceans ... are the "black smokers," volcanic gases and metal enriched hot brines emerging in isolated places from the floor and giving a home to a highly distinctive fauna adapted to this unusual environment. The only possible ancient example I know is in the Upper Jurassic of the French Alps.[34]

If the majority of the rocks (since the Cambrian) are part of a SERGA, they we would expect few if any hydrothermal vents. The lack of hydrothermal vents supports SERGA.

In the Mowry and Aspen Shales (mostly in WY and MT) we find calcareous concretions, around six feet in diameter, with an amazingly diverse menagerie of fossils. These include carbonized wood, marine reptiles, pterodactyls, fish and ammonites.[35] We don't normally think of ammonites and pterodactyls as part of the same ecological niche. What cataclysm could have brought these various organisms together? Also, if we dated the wood by radiocarbon dating would we get a date of just thousands of years? If so, the early Cretaceous date (~140M) of these shales must be wrong.

In describing Bermuda, Darwin talks about,

> ... sand drifted by the winds and agglutinated together ... there occur in one place five or six layers of red earth ... including stones too heavy for the wind to have moved. ... Mr. Nelson [Lt. Nelson of the Royal Engineers] attributes the origin of these several layers, with their embedded stones to <u>violent catastrophes</u> ... further investigation has generally succeeded in explaining such phenomena by <u>simpler means</u>.[36]

When the evidence for catastrophism was starring Darwin in the face, he denied it.   He was blinded by his Lyellian gradualistic PreSuppositions.   Another example of Darwin refusing catastrophism is when on February 20, 1835 a massive earthquake hit Valdivia, Chile. Darwin was right there on the shore near Valdivia!   The quake has been estimated as magnitude 8.5 and virtually all the buildings in Valdivia were destroyed. Rocks lined with recent marine shells were elevated above the tide. The island of Santa Maria was raised by an average of 9 feet!! Parts of the nearby island of Quiriquina had risen a few feet due to the earthquake.[37, 38]

Large Lava Layers

Large igneous provinces (LIP's) are formed catastrophically and create high-volume accumulations of volcanic and intrusive igneous rock. A number of these LIP's are so large that if taken together would have covered Texas 27 times! Additionally, if we take the cumulative volumes of this group of LIP's the total volume would form a cube 171 miles wide!![39]

Large Igneous Province	Area M km^2	Volume M km^3
Columbia River Basalt	0.16	0.18
Afro-Arabia	0.60	0.35
North Atlantic	1.30	6.60
Deccan Traps (India)	0.65	0.75
Ontong-Java	1.86	8.40
Karoo-Ferrar	0.18	0.30
Central Atlantic	11.00	2.50
Siberian Traps	2.70	1.45
Emeishan (China)	0.25	0.30
Total	**18.70**	**20.83**

Clearly, in the past, volcanic activity was on a much more massive scale than today. Can volcanic mountains form fast? The Jorullo volcano in Mexico was born on September 29, 1759 and erupted for 15 years straight. Jorullo grew 820 ft (250 m) from the ground in the first six weeks.[40]

Under the proper conditions, basalt formations can solidify quickly. As Herbert Huppert maintains,

> When basaltic magma is emplaced into continental crust, melting and generation of granitic magma can occur. ... Our calculations applied to basaltic sills in hot crust predict that sills from 10-1500m thick require only 2-200 years to solidify ...[41]

So the time factor in the formation of igneous rock is consistent with YES and a SERGA.

Tracking Pterosaurs (& Birds)

Ricardo Melchor from the Universidad Nacional de La Pampa in Argentina, reports on the discovery of bird tracks from the late Triassic, "It's significant because we find footprints with morphology identical to modern birds in rocks that predate, by 55 million years, the first record of true Aves."[42] These footprints predate, by 55 million years, the oldest bird body fossils. Late Triassic age bird tracks have also been found in Africa. Birds have an advantage in fleeing from a global paroxysm. Could this be why we find their tracks before their body fossils? This is consistent with a SERGA.

The earliest pterosaur tracks may be found in the Late Jurassic.[43] Most pterosaur body fossils are found in the Cretaceous.[44] This is exactly what we would expect if this orb has experienced an overwhelming global natural disaster. Because of pterosaurs' ability to flee danger, their footprints would appear in the fossil record before their bones.

The Cambrian Next Door

Are there indicators that geologic events that supposedly occurred millions of years ago were actually part of human history? The Vale of Kashmir lies in the high mountainous between India and Pakistan. The Hindu *Nilamata Purana* (c. between 550 AD and 700 AD) describes the Valley of Kashmir as once having an enormous lake. The legend tells of the water leaving the lake all at once, in a violent breaching event which created rush of water with gigantic waves within human memory.[45]

J.A. Khan et al state, "Geological evidences shows that Kashmir was once a glacial lake which occupied the whole area of the valley. The lake found its outlet in the South-West of Srinagar a few million years back."[46] How could these legends have survived for millions of years? Or did this lake overflow catastrophe happen just a few thousand years ago? Was the end of the Ice Age all that long ago? Could our dating schemes be flawed? Could Cambrian rocks actually be only thousands of years old? Is the Cambrian Period closer than we think?

ONE Ice Age

Paleontologist Jean Louis Agassiz (d. 1873) held to the stasis of Essential Types of Life (ETL's, Ch. 3) and the catastrophism of Cuvier. He introduced the idea of a great Ice Age.[47]

Multiple ice ages would seem to contradict YES and a SERGA. Karl Alfred von Zittel, Professor of Geology and Paleontology at the University of Munich (writing in 1901) said,

> Numerous observations in different areas have testified to the frequent oscillations of the glaciers during the Ice Age.... Holst in Norway, Upham and Wright in North America, and many other authorities recognize only one Ice Age, marked by occasional seasonal or periodic variations of no great significance in the dimensions of the glaciers and inland ice. It is still more doubtful whether geologists have been right in supposing that several Ice Ages occurred during geological epochs ...[48]

In a more recent publication (1994), Robert Young et al, writing in *Geology*, conclude that, "... it is likely that large areas of higher elevation in Alberta, south and west of Edmonton, only underwent the late Wisconsin event, not multiple glaciations."[49] So, at least for southwest Alberta, mainstream geology agrees to only ONE Ice Age.

What about the origin of the Ice Age? Volcanism in the past has been much more severe than recent times and could have spawned an Ice Age. According to Fred Warshofsky,

> Increased volcanic activity ... is now believed to be directly linked to the creation of an ice age. ... The dust veils produced by volcanoes reduce the amount of sunlight reaching the earth. Volcanic dust also acts as nuclei around which ice crystals form, producing clouds, which further cut the amount of sunlight that reaches the earth.[50]

Geologist Jon Erickson (*Quakes, Eruptions and Other Geologic Cataclysms*) provides a key factor that may have initiated the Ice Age. He boldly comments on the impact of asteroids:

> Large meteorite impacts can create so much disturbance in the Earth's crust that volcanoes and earthquakes could become active in zones of weakness. A massive impact in the Amirante Basin, 300 miles northeast of Madagascar, might have triggered India's great flood basalts, known as the Deccan Traps…Quartz grains shocked by high pressures generated by a large meteorite impact found lying just beneath the immense lava flows might be linked to the impact. … <u>A single large meteorite impact</u> or a massive meteorite shower would eject <u>tremendous amounts of debris into the global atmosphere.</u> This would block out the Sun for many months or years, <u>possibly bringing down surface temperatures significantly to initiate glaciation.</u>[51]

From 1811 to 1812, Tambora ejected 220M metric tons of ash into the atmosphere. The average temperature in central England dropped by 4.5° F.[52]

Under these conditions, can glaciers grow fast? During the Little Ice Age (1550 – 1850 AD), glaciers in Norway advanced by several kilometers each year. Sea ice reached Faroe Islands (Denmark), just 250 miles north of Britain![53] Iceland's Brúarjökull glacier advanced 5 miles in one year![54]

a.f. nsidc.org

During the period from 2007 to 2009 the giant ice sheet at the North Pole grew rapidly. In like manner, Warshofsky summarizes how glaciers grow,

> Glaciers literally feed themselves. Moist air from the oceans drifts over the continent and bumps into the advancing edge of ice. The air rises to pass over the ice mass, expanding and cooling as it goes. The dropping temperature reduces the air's ability to hold moisture. Clouds form, grow heavier and let their moisture fall as snow upon the already immense glacier.[55]

a.f. p38assn.org

Snow and ice can indeed accumulate rapidly. *The Lost Squadron* documents warplanes abandoned on Greenland during World War II. Their burial shows that the Ice Age did not take long ages to appear. In just fifty years (1942-1992) about 79 feet of snow and 200 feet of ice formed above the planes. Thus, it does not multiple thousands of years for glaciers to form. One P-38 is called "Glacier Girl."[56]

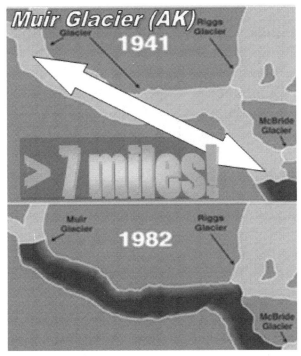

a.f. wikipedia

Muir Glacier in Alaska retreated over 7 miles in just 41 years! So, it did not take vast ages for the Ice Age to end. Warshofsky explained that the Ice Age could have closed quickly,

> Did it take multiple thousands of years for the Ice Age to end? A study by the Geological Survey of Canada concluded that a mile thick ice sheet that existed thousands of years ago slipped into Hudson Bay in less than two centuries. For many ice sheets there is an underlying lubricating slush created from the heat of the earth that aids the ice sheet to glide into the ocean.[57]

a.f. wikipedia

There have been several *ICE AGE* films, but only <u>one</u> Ice Age. In the book *Earth's Glacial Record* (Cambridge Univ. Press) we find that some features indicative of glacial activity, may actually be due to mass flows:

> Schermerhorn argued that tectonic differentiation involving complementary uplift and subsidence set the scene for deposition of many Neoproterozoic 'glacial' successions. Schermerhorn's arguments were largely ignored by geologists who felt that he had largely written off *all* Neoproterozoic tillites in favor of non-glacial, tectonically generated mass flows. … Schermerhorn had earlier argued that many supposed tillites were non-glacial mass flows …[58]

Earth is Obviously Old

One looks at the Grand Canyon and many would say "obviously it took millions of years to form those layers and erode out the canyon." But consider the Seven Devils Gorge (Hell's Canyon, Snake River) which is between Oregon and Idaho. It's the deepest canyon in North America (sorry Grand Canyon). Seven high peaks form a semicircle. These peaks are called the Seven Devils Mountains. The following legend has been corroborated and details were added by Caleb Whitman, a Nez Perce on the Umatilla Reservation (August 1950):

… Coyote called together all the animals with claws … to dig <u>seven deep holes</u>. Then Coyote filled each hole with a <u>reddish-yellow liquid</u>. His good friend Fox helped him <u>keep the liquid boiling</u> by dropping hot rocks into it. Soon the time came for the giants' journey eastward. … Down, down, down the seven giants went into the <u>seven deep holes of boiling liquid</u>. They struggled and struggled to get out, but the holes were very deep. They fumed and roared and splashed. As they struggled, they scattered the reddish liquid around them <u>as far as a man can travel in a day</u>. … Coyote said to the seven giants, "I will punish you even more by changing you into <u>seven mountains</u>. I will make you very high, so that everyone can see you. You will stand here forever to remind people that punishment comes from wrongdoing." … Coyote caused the <u>seven giants to grow taller</u>, and then he changed them into <u>seven mountain peaks</u>. He struck the Earth a hard blow and so opened up a deep canyon at the feet of the giant peaks. <u>Today the mountain peaks are called the Seven Devils. The deep gorge at their feet is known as Hell's Canyon of the Snake River</u>. And the <u>copper</u> scattered by the splashing of the seven giants is <u>still being mined</u>.[59]

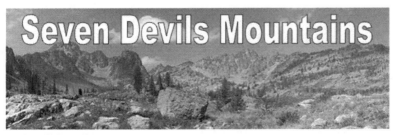

a.f. wikipedia

So we have an indication that this canyon formed within the memory of man and not tens of thousands of years ago as establishment geology maintains. Would the Native traditions have remained unmodified for 10K or 20K years?

3) "Causal biostratigraphy" by Valentin Krassilov, *Lethaia*, Vol. 7, no. 3, pp. 173-179, p. 173, emphasis added.
http://www.paleobotany.ru/PDF/1970-1979/Krassilov_1974_casual_biostr.pdf

4) Ibid., p. 174, emphasis added.

5) "The Importance of Depostional Sorting to the Biostratigraphy of Plant Megafossils" by Robert Spicer in *Biostratigraphy of Fossil Plants* ed. by David Dilcher and Thomas Taylor (Dowden, Hutchinson and Ross, Stroudsburg, PA, 1980), pp. 171, 181, emphasis added.

6) "Consider the incompleteness of the geological record" by Tjeerd van Andel, *Nature*, Vol. 294, Is. 5840, pp. 397-398 (Dec. 3, 1981), p. 397, emphasis added.

7) *The New Catastrophism* by Derek Ager (Cambridge University Press, 1993), p. 14.

8) *Evolution, Creationism, And Other Modern Myths* by Vine Deloria Jr. (Fulcrum Pub., Golden, CO, 2002), p. 92.

9) "Mountain-Building Chronology and Nature of Geologic Time Scale" by Edmund Spieker, *American Association of Petroleum Geologists' Bulletin*, 40:1769-1815, Aug. 1956, p. 1815.

10) *The New Catastrophism* by Derek Ager (Cambridge University Press, 1993), p. 120, emphasis added.

11) *This Land: A Guide to Central National Forests* by Robert Mohlenbrock (Univ. of California Press, Berkeley, 2006), pp. 215, 216.

12) "Reinterpretation of the depositional environment of the Yellowstone 'fossil forests'" by William J. Fritz, *Geology*, Vol. 8, July 1980, pp. 309-313, pp. 309, 312.

13) *The New Catastrophism* by Derek Ager (Cambridge University Press, 1993), p. 42.

14) *Principles of Stratigraphy* by Carl Dunbar and John Rodgers (John Wiley & Sons, NYC, 1957), p. 121, emphasis added.

15) Ibid., p. 127, emphasis added.

16) Ibid., p. 127, emphasis added.

17) "Paraconformities" by Norman D. Newell in Essays in Paleontology and Stratigraphy ed. by Curt Teichert and Ellis Yochelson (Univ. of Kansas Press, Lawrence, KS, 1967), pp. 349-367, p. 355, emphasis added.

18) *Geologic Time* by Don Eicher (Prentice-Hall, Englewood Cliffs, NJ, 1968), p. 35.

19) "In Our Time – Debate"
http://www.bbc.co.uk/radio4/history/inourtime/inourtime_comments_ageing_earth.shtml

20) *The Face of the Earth* by Eduard Suess (Clarendon Press, Oxford, 1906), Vol. 2, trans. by Hertha Sollas, p. 543, emphasis added.

21) *Summary Report of the Operations of the Operations of the Geological Survey* [of Canada] *for the Year 1886* by Alfred Selwyn (Dawson Brothers, Montreal, 1887), p. 8A, emphasis added.
http://books.google.com/books?id=8BkeAAAAYAAJ

22) "A Tidal Model of Carboniferous Konservat-Lagerstätten Formation" by Howard Feldman, Allen Archer, Erik Kvale, Christopher Cunningham, Christopher Maples and Ronald West, *PALAIOS*, Vol. 8, No. 5 (Oct., 1993), pp. 485-498, p. 485, emphasis added.
http://www.jstor.org/pss/3515022

23) "Spontaneous stratification in granular mixtures" by Hernán A. Makse, Shlomo Havlin, Peter R. King and H. Eugene Stanley, *Nature*, Vol. 386 (1997), pp. 379 - 382, p. 379, emphasis added.
http://www.nature.com/nature/journal/v386/n6623/abs/386379a0.html

24) "Kenneth J. Hsu: Catastrophes, Dinosaurs and Evolution" [Review] by Emerson Thomas McMullen (1999).
https://sites.google.com/a/georgiasouthern.edu/etmcmull/kenneth-j-hsu-catastrophes-dinosaurs-and-evolution

25) "Polystrate fossil"
http://en.wikipedia.org/wiki/Polystrate_fossil

26) "Chapter XXIII, The Coal Or Carboniferous Group," emphasis added.
http://geology.com/publications/lyell/ch23.shtml

27) *The New Catastrophism* by Derek Ager (Cambridge University Press, 1993), p. 47.

28) "Sandstone Pipes of the Laguna Area, New Mexico" by John Schlee in *Unknown Earth* by William Corliss (The Source Project, Glen Arm, MD, 1980), pp. 16, 17.

29) *The New Catastrophism* by Derek Ager (Cambridge University Press, 1993), p. 49, emphasis added.

30) "Origin of a thick, redeposited carbonate bed in Eocene turbidites of the Hecho Group, south-central Pyrenees, Spain" by D. R. Johns, E. Mutti, J. Rosell and M. Séguret, *Geology*, Vol. 9, Apr. 1981, pp. 161-164, p. 161.
http://geology.gsapubs.org/content/9/4/161.abstract

31) *Geologic Time* by Don Eicher (Prentice-Hall, Englewood Cliffs, NJ, 1968), p. 72.

32) *The New Catastrophism* by Derek Ager (Cambridge University Press, 1993), pp. 107, 108.

33) Ibid., p. 110, emphasis added.

34) Ibid., p. 114.

35) "Ammonite Accumulations in the Cretaceous Mowry and Aspen Shales" by John Reeside in *Unknown Earth* by William Corliss (The Source Project, Glen Arm, MD, 1980), p. 682.

36) Quoted in *The New Catastrophism* by Derek Ager (Cambridge University Press, 1993), pp. 118, 119, emphasis added.

37) "Outline of Darwin's Beagle voyage"
http://www.aboutdarwin.com/timeline/time_04.html

38) "What 1835 Chile quake taught Darwin" by John van Wyhe
http://www.cnn.com/2010/OPINION/03/01/vanwyhe.quake.chile.darwin/index.html

39) "Large igneous province"
http://en.wikipedia.org/wiki/Large_igneous_province

40) "Jorullo, Mexico"

http://volcano.oregonstate.edu/vwdocs/volc_images/north_america/mexico/jorullo.html

41) "The fluid dynamics of crustal melting by injection of basaltic sills" by Herbert Huppert and Stephen Sparks, *Transactions of the Royal Society of Edinburgh: Earth Science*, Vol. 79, pp. 237-243, p. 237.

42) "Footprints step back bird origins?" by Salma Monani, http://www.geotimes.org/june02/WebExtra0627.html

43) "Pterosaur swim tracks and other ichnological evidence of behaviour and ecology" by Martin Lockley and Joanna Wright, http://sp.lyellcollection.org/content/217/1/297.abstract

44) "List of pterosaur genera" http://en.wikipedia.org/wiki/List_of_pterosaurs

45) "Kashmir" by David Mathisen http://mathisencorollary.blogspot.com/2011/06/kashmir.html

46) "Exploring present status of Hydrochemistry and Sediment chemistry of Dal Lake, Kashmir and effect of anthropogenic, Disturbances on it" by J.A. Khan, R.S. Gavali and Y.S. Shouche, *Indian Journal Innovations and Developments*, Vol.1, No.7 (July 2012), p. 556, emphasis added. http://www.academia.edu/2692062/Exploring_present_status_of_Hydrochemistry_and_Sediment_chemistry_of_Dal_Lake_Kashmir_and_effect_of_anthropogenic_Disturbances_on_it

47) "Jean Louis Rodolphe Agassiz" by James S. Aber http://academic.emporia.edu/aberjame/histgeol/agassiz/agassiz.htm

48) *History of Geology and Paleontology to the End of the Nineteenth Century* by Karl Alfred von Zittel (Walter Scott, London, 1901), trans. by Maria Ogilvie-Gordon, p. 235, emphasis added.

49) "A single, late Wisconsin, Laurentide glaciation, Edmonton area and southwestern Alberta" by Robert Young, James Burns, Derald Smith, David Arnold and Bruce Rains, *Geology*, Vol. 22 (1994), no. 8, pp. 683-686, p. 683, emphasis added. http://geology.gsapubs.org/content/22/8/683.abstract

50) *Doomsday: the Science of Catastrophe* by Fred Warshofsky (Reader's Digest Press, NYC, 1977), pp. 167, 168, emphasis added.

51) *Quakes, Eruptions and Other Geologic Cataclysms* (*Revealing the Earth's Hazards*, Rev. Ed.) by Jon Erickson (Checkmark Books, NYC, 2001), p. 244, emphasis added.

52) *Doomsday: the Science of Catastrophe* by Fred Warshofsky (Reader's Digest Press, NYC, 1977), p. 183.

53) Ibid., p. 174.

54) *Quakes, Eruptions and Other Geologic Cataclysms* (*Revealing the Earth's Hazards*, Rev. Ed.) by Jon Erickson (Checkmark Books, NYC, 2001), p. 212.

55) *Doomsday: the Science of Catastrophe* by Fred Warshofsky (Reader's Digest Press, NYC, 1977), p. 177, emphasis added.

56) "Glacier Girl Recovery" http://p38assn.org/glaciergirl/recovery.htm

57) *Doomsday: the Science of Catastrophe* by Fred Warshofsky (Reader's Digest Press, NYC, 1977), p. 191.

58) *Earth's Glacial Record* by M. Deynoux, J. M. G. Miller, E. W. Domack (Cambridge University Press, 1994), pp. 9, 10.

59) "Chapter 7 Literature: The Seven Devils Mountains - Nez Perce," emphasis added.
http://www.pacificnorthwestjourneys.org/year2/supplements/lit.cfm?chid=1
7

60) "Novarupta"
http://en.wikipedia.org/wiki/Novarupta

61) *Sahara* by John Julius Norwich (Prentice Hall, 1968), p. 103

62) "Sahara"
http://en.wikipedia.org/wiki/Sahara_desert

63) *Timeline* by Michael Crichton (Ballantine Books, NYC, 1999), pp. 302, 303.

64) *The Age Of The World* by Francis Haber (Johns Hopkins Press, Baltimore, MD, 1959), p. 197.

Chapter 6
Dub Step Earth – Pangea and RPM's

Fly the great big sky see the great big sea
Kick through continents bustin' boundaries ...
Roam if you want to, roam around the world
Roam if you want to, without wings without wheels
 - B-52's

With other relics of "a former World"
When this World shall be former, underground
Thrown topsy-turvy, twisted, crisped, and curled
Baked, fried, or burnt, turned inside-out, or drowned
 - Lord Byron

Continental Drift or RUN?

Since the 1960's plate tectonics has become generally accepted
as an accurate model of shifting continents. Richard Moody Jr.
(Masters in Geology, SUNY Albany) believes that plates may
move as fast as 1m/year.   A plate movement rate of 20cm/yr.
is considered "fast" by most geologists. Moody explains, "The
entire plate can contribute to that heat [from shear] as friction
and shed the lower melting fractions, e.g. iron rich olivine
which may form a Teflon-like character to the plate boundary a
few centimeters thick."[1]

Could this facilitate Rapid Plate Movements (RPM)?   Could
the plates move much faster than 1m/year?

Moody supports RPM:

> ... for brief periods of time, plates might achieve
> velocities 10-100 times their steady-state displacement.
> ... Slab pull becomes enormous and the only limiting

variable as far as the rapidity of subduction is how the down going slab gets rid of all the subducted material.[2]

So plates may move as fast as 20m/yr. (20cm * 100). Could they move even faster?

a.f. *The Science of Evolution*

Alfred Lothar Wegener (Ph.D. Astronomy, University of Berlin) was the Father of Continental Drift and held that North America and Europe were moving apart from each other at 2.5 m each year!![3]

Venus and RPM's

James W. Head III, in a paper in *Nature*, proposes that there was a great flood on Venus and that it has experienced "rapid plate tectonics."[4] Could the same RPM's have occurred on Earth?

Mark Bullock and Head (planetary geologist at Brown Univ.) have used simulation to support the catastrophic resurfacing hypothesis for Venus.[5] If the plates move fast on Venus, what about this planet?

The Hapgood Connection

Charles Hapgood (*Maps of the Ancient Sea Kings*) advocated catastrophic pole shifts for the Earth. Hapgood's work was the basis for the film *2012*. Even Stephen Jay Gould suggested that this theory deserves a fair hearing and should not be dismissed out of hand.[6] Can the pole shift idea be rehabilitated by combining it with plate tectonics? YES it can - we call this Rapid Plate Movements (RPM's).

Albert Einstein wrote the Foreword to Hapgood's book,

> A great many empirical data indicate that at each point of the earth's surface that has been carefully studied, <u>many climatic changes have taken place, apparently quite suddenly</u>. This, according to Mr. Hapgood, is explicable if the virtually <u>rigid outer crust of the earth</u> undergoes, from time to time, <u>extensive displacement</u> over the viscous, plastic, possibly fluid inner layers.[7]

Kirtley Mather, Harvard Geology Professor and former President of the American Association for the Advancement of Science (AAAS), said in 1959 that Hapgood's pole shift theory is "worthy of careful study and appraisal."[8]

Albert Einstein himself thought Hapgood's proposal deserved real consideration,

> [Hapgood's data] … is explicable if the virtually rigid outer crust of the earth undergoes, from time to time, extensive displacement over the viscous, plastic, possibly fluid inner layers. … I think that this rather astonishing, even fascinating, idea deserves the serious attention of anyone who concerns himself with the theory of the earth's development.[9]

Note how similar this is to RPM.

Albert Einstein wrote <u>ten</u> letters to Hapgood. Einstein told Hapgood, "I find your arguments very impressive and have the impression that your hypothesis is correct. One can hardly doubt that significant shifts of the crust have taken place repeatedly and within a short time." Hapgood and Einstein actually met in January of 1955.[10] Hapgood held that the last shift of the lithosphere took place at the end of the Ice Age.[11]

Consider the following remarks of Hapgood in the context of RPM's:

> … the viscosity of the asthenosphere would be reduced to its lowest point by the fluid wave-guide layer, and so the lithosphere would in effect be borne along on a stream flowing in a liquid … <u>Friction would be minimized</u> … the zone of easy shear in the wave-guide layer opens up the <u>possibility of extremely rapid movements of the earth's outer shell</u>.[12]

Wegener assumed that Europe and North America were quite close to each other during the ice age.[13] Given the mainstream date that the ice age ended 10K years ago, that comes out to plate movements of 1,500 feet each year! RPM indeed.

Rapid Vertical Movement

There is evidence of rapid vertical uplift on the mountains of Greece. The geologist P. Negris found signs of ancient beaches as high as 1,700 feet elevation! Could these remains have survived for so long unless these mountains were raised recently?[14]

William H. Hobbs provides another example: "Upon the coast of Southern California may be found all the features of wave-cut shores now in perfect preservation, and in some cases as much as fifteen hundred feet above the level of the sea."[15]

Lake Titicaca, between Peru and Bolivia, contains seahorses! It also contains the marine genus *Allorchestes*. Could this indicate that lake was connected to the ocean in the not too distant past? There are also saltwater mollusks. Lake Titicaca is over 12,000 feet high, yet nearby Tiahuanaco with its magnificent ruins show that it was built near a lake![16] This makes sense if the Andes rose rapidly.

Hapgood provides further evidence for the rapid rise of the Andes:

> Remarkable confirmation of the immensity of this uplift is represented by the ancient agricultural stone terraces surrounding the Titicaca basin. These structures, belonging to some bygone civilization, occur at altitudes far too high to support the growth of cops for which they were originally built. Some rise to 15,000 feet above sea level, or about 2,500 feet above the ruins of Tiahuanaco, and on Mt. Illimani they occur up to 18,400 feet above sea level; that is, above the line of eternal snow.[17]

The Himalayan mountains are commonly thought to have arisen 50M years ago. However, recently in 2008 Yang Wang et al of Florida State University found thick layers of ancient lake sediment filled with plant, fish and animal fossils typical of far lower elevations and warmer, wetter climates. Paleomagnetic studies dated these features as young as 2M years!! *Science Daily* reports, "The new evidence calls into question the validity of methods commonly used by scientists to reconstruct the past elevations of the region."[18]

Derek Ager states, "… there were many phases of <u>mountain building movements</u>, each of them of short duration and therefore what might be called <u>episodic catastrophes</u>."[19]

a.f. phys.org

In October 2013 a 7.1-magnitide quake hit the Philippines on the central island of Bohol. This created a wall of rock 10 feet tall when the ground moved vertically. The wall of rock runs for at least 3 miles![20] Could this be a clue that RPM is true? Rapid movement of the Earth is possible.

Richard Oldham (d. 1936) was a British geologist who was the first to distinguish P-waves, S-waves and surface waves. His report on the 1897 Assam earthquake described the Chedrang fault, with uplift up to 35 feet and ground movements greater than Earth's own gravitational acceleration![21]

Displacement Bay, Yakutat Bay, Alaska in 1899 experienced the largest vertical surface displacement caused by a single event. Earthquakes thrust underwater rocks 47 feet above the sea.[22]

Writing in *Geology: a survey of earth science,* Edgar Spencer states, "<u>Periods of mountain formation</u> mark breaks between some of the major divisions of geologic time. <u>Many have envisioned these mountains as rising rapidly, in a matter of a few hundreds of years</u>."[23] Mountains do not take multiple eons to form. This fits perfectly with RPM.

Cuba exhibits rapid uplift in recent times. Raised shorelines are observed along with a dozen terraces as high as 1000 feet above sea level. If this uplift had taken millions of years these terraces would have been eroded away.[24]

Dr. William Van Dorn gave the following account of the great earthquake of 28 March 1964 on southeastern Alaska, and its tsunami:

> The volume of water displaced defies imagination. It involved a dislocation averaging 6 feet vertically over 100,000 square miles - thrice the size of Florida! About half this area was on land, and subsided; the other half, which included the entire 100-mile wide shelf bordering the Gulf of Alaska, was bulged upward - in some places as much as 50 feet.[25]

British Pleistocene geologist, John Kaye Charlesworth claimed,

> The Pleistocene indeed witnessed Earth-movements on a considerable, even catastrophic scale. There is evidence that it created mountains and ocean deeps of a size previously unequalled ... Earth movements elevated the Caucasus ... [since] the second stage of the Ice Age is estimated at 3900 feet... and raised Lake Baikal region ... and Central Asia by 6700 feet. Similar changes took place around the Pacific and in North China the uplift was estimated at 10,000 feet.[26]

Rapid Horizontal Movement

In order to deal with the difficulties arising from tidal friction in the Earth-Moon system, R.A. Lyttleton proposed that the Earth is contracting. He theorizes that this planet underwent a reduction in radius of 70 km in just a matter of minutes![27] If Lyttleton was given a hearing for this view, should not RPM proponents be allowed a fair chance to argue their case as well?

In 1923 in Kanto, Japan (the Great Tokyo Earthquake) a 7.9 magnitude earthquake produced nearly 6 feet of vertical uplift on the north shore of Sagami Bay and horizontal displacements of as large as 15 feet![28]

In 2005, a magnitude 7.6 earthquake shook the Kashmir region of Pakistan and a 5m shift occurred. This movement was about 50 degrees from horizontal – that is, the movement was both vertical and horizontal.[29]

In December of 2004 the Nicobar Islands not only experienced a devastating tsunami, but the islands themselves were moved by as much as 100 feet by the earthquake! Could this be a clue to RPM's?[30]

a.f. geology.ohio-state.edu

At the same time as the San Francisco earthquake of 1906, one road that crossed the San Andreas Fault was moved 21 feet horizontally![31] Could a sustained period of earthquakes move the earth 200 feet or even two miles?

The 2004 Indian Ocean earthquake centered off of Sumatra demonstrated RPM's. There was 33 feet movement laterally and 13-16 feet vertically along the fault line![32] Spectacular and catastrophic earth movements do occur. If such movements were continuous and widespread this would support RPM.

Derek Ager provides a fascinating observation about the African Rift Valley,

Where the plates are really active, as in the Rift Valleys of East Africa, the movements are very obvious. In the Afar Triangle of north-east Africa, cracks can be seen in the ground where the <u>two sides of the Great Rift are still moving apart. It is said that if you put a stick across one of these cracks in the morning, by the evening it will have fallen in.</u>[33]

Could this be a remnant of RPM's?

The Indian-Australian plate is diving into the Tonga Trench (near Fiji) at nine inches per year![34] Seafloor spreading is up to <u>ten times faster</u> for the East Pacific Rise compared to the Mid-Atlantic Ridge.[35] A model that appeared in *Geology* showed a seafloor spreading rate of 92 mm per year in the Early Cretaceous – much faster than current rates.[36]

Thermal Runaway and RPM's

Robert D. Ballard, who discovered the Titanic and the Bismarck, details the thermal runaway effect:

> … a rough spot on the underside of the plate generates heat as it rubs across the asthenosphere below. Small pockets of magma formed by the heat float upward. Again pressure drops, triggering more melting. This feedback leads to a self-perpetuating "thermal runaway" …[37]

That is, more magma creates more magma. Could this process fuel RPM?

Yuri Fialko and Yakov Khazan (Scripps Institution of Oceanography) conclude that,

> After the onset of thermal runaway, viscous resistance of the melt layer becomes negligible, and slip accelerates to a limiting velocity corresponding to the

elastodynamic rupture with a nearly complete stress drop ...[38]

Thus, thermal runaway (unlimited acceleration and heating of the melt layer) could cause plate tectonics to speed up.

Hans-Peter Bunge et al admit,

> Lithospheric plate motions at the Earth's surface result from thermal convection in the mantle. ... The plates themselves result from high rock strength and brittle failure at low temperature near the surface. In the deeper mantle, elevated pressure may increase the effective viscosity by orders of magnitude. ...These effects of depth-dependent viscosity may be comparable to the effects of the plates themselves.[39]

Could this imply that Earth's tectonic plates may have moved 100 or 1000 times faster in the past than current rates? Go RPM!!

Magnetic Reversals – it's about TIME (& Space)

Don't magnetic reversals along the Mid-Ocean Ridges demonstrate that plate movements are slow? In 1989, R.S. Coe and M. Prévot published "Evidence suggesting extremely rapid field variation during a geomagnetic reversal."[40] And in 1995 they published similar evidence in the journal *Nature*.[41] Amazingly, 90 degrees of reversal can occur in just 15 days!!

Can meteors and comets lead to a magnetic reversal? Writing in *Geophysical Research Letters*, Richard Muller and Donald Morris contend that, "The impact of a large extraterrestrial object on the Earth can produce a geomagnetic reversal ..."[42] Thus, reversals may happen suddenly. This supports RPM's.

Who Supports RPM?

Writing in the *Journal of Physics*, Peter Warlow (*The Reversing Earth*) stated,

> The enigma of geomagnetic reversals and their apparent link with other phenomena, such as faunal extinctions, is shown to be explicable by treating these reversals as a relative rather than an absolute effect. Instead of reversing the magnetic field, it is suggested that a reversal of the Earth itself ... It is shown that a wide variety of data is compatible with this hypothesis, not only from modern geological and related investigations, but also from astronomy and from ancient sources.[43]

In other words, Warlow advocates quick magnetic reversals. Of course, the Earth flipping would be a major catastrophe and cause massive flooding. Is this theory any more extreme than RPM?

Fred Warshofsky explicitly supports RPM:

> There are other pieces of evidence that indicate <u>shifts in the position of the continents that cannot be described as slow-motion catastrophe</u> [e.g. origin of the Himalayas]. To account for these, one requires <u>sudden, full-blown, planet-shaking cataclysms</u>.[44]

In 1658, R.P. François Placet proposed that the Old and New Worlds separated catastrophically.[45] So RPM is no new idea. Antonio Snider-Pellegrini (d. 1885) was a French geographer who proposed continental drift before Wegener. He based this theory on the fact that he had found plant fossils in both Europe and the United States that were identical. He found matching fossils on all of the continents. He held that the breakup of the former giant supercontinent was catastrophic.[46]

What Event Initiated RPM?

Yukio Isozaki, writing in the *Journal of Asian Earth Sciences*, posits that a catastrophic mantle superplume created a Large Igneous Province (LIP) and split the supercontinent Pangea.[47] Luis and Walter Alvarez proposed in 1980 that the impact of a large asteroid at the Cretaceous-Tertiary boundary (KT boundary) caused the demise of the dinosaurs. Richard Huggett points to, "… major impacts on continents as possible triggers of tectonic or geomagnetic changes. … It has been argued that bombardment may play a basic role in plate tectonics, in geomagnetic reversals … earthquakes [and] volcanism …"[48]

Rapid Plate Movements (RPM) could have energized the Singular Epoch of Rapid Geologic Activity (SERGA). Please carefully consider the evidence presented in this chapter. The science is in — RPM is **real**. Clearly, RPM and Young Earth Science (YES) fit together like hand and glove. If most geologic events occurred rapidly, how does this affect the Time of Man versus the Time of Trilobites? We survey this topic next.

Notes:
1) "Beyond Plate Tectonics: 'Plate' Dynamics" by Richard Moody Jr., *Infinite Energy*, Issue 74, 2007, p. 3.
http://www.infinite-energy.com/images/pdfs/moody.pdf
2) Ibid., p. 7.
3) "Alfred Wegener (1880-1930)"
http://www.ucmp.berkeley.edu/history/wegener.html

4) "Venus After the Flood" by James W. Head III, *Nature*, Vol. 372 (1994), Issue 6508, p. 729.

5) "Venus Unveiled" by David Grinspoon, *The Sciences*, Jul./Aug. 1993, pp. 20-26, p. 25.

6) *Catastrophism* by Richard Huggett (Verso, London, 1997), p. 120.

7) *The Path of the Pole* by Charles Hapgood (Chilton Book Co., Philadelphia, PA, 1970), 2 ed. of *Earth's Shifting Crust* (1958), p. xiv, emphasis added.

8) Ibid., p. xii.

9) Ibid., pp. xiv, xv.

10) "The Einstein-Hapgood Papers" by Marc Bergvelt
http://www.oocities.org/athens/troy/6396/lightfall0341.htm

11) *The Path of the Pole* by Charles Hapgood (Chilton Book Co., Philadelphia, PA, 1970), 2 ed. of *Earth's Shifting Crust* (1958), p. xvi.

12) Ibid., p. 43, emphasis added.

13) Ibid., p. 25.

14) Ibid., p. 238.

15) quoted in Ibid., p. 238.

16) Ibid., pp. 281-283.

17) Ibid., p. 285.

18) "Fossils Found In Tibet Revise History Of Elevation, Climate"
http://www.sciencedaily.com/releases/2008/06/080611144021.htm

19) *The New Catastrophism* by Derek Ager (Cambridge University Press, 1993), p. xvi, emphasis added.

20) "Philippine earthquake creates miles-long rocky wall"
http://phys.org/news/2013-10-philippine-earthquake-miles-long-rocky-wall.html

21) "Richard Dixon Oldham"
http://en.wikipedia.org/wiki/Richard_Dixon_Oldham

22) "Earthquakes: Case Studies"
http://library.thinkquest.org/C003603/english/earthquakes/casestudies.shtml

23) *Geology: a survey of earth science* by Edgar Winston Spencer (Crowell, 1965), p.427, emphasis added.

24) "Raised Shore-Lines in Cape Maysi, Cuba" by Oscar Hershey in *Unknown Earth* by William Corliss (The Source Project, Glen Arm, MD, 1980), pp. 358, 359.

25) "Oceanography: special waves" by J. Floor Anthoni
http://www.seafriends.org.nz/oceano/waves2.htm

26) *The Quaternary Era*, Vol. 1 by John Kaye Charlesworth (Edward Arnold, 1957), pp.604-605, emphasis added.

27) *The Earth and Its Mountains* by R.A. Lyttleton (John Wiley & Sons, Chichester, UK, 1982), p. xvii.

28) "Earthquakes with 1,000 or More Deaths since 1900"
http://earthquake.usgs.gov/earthquakes/world/world_deaths.php

29) "The Kashmir Earthquake of October 8, 2005: Impacts in Pakistan"
EERI Special Earthquake Report — February 2006
https://www.eeri.org/lfe/pdf/kashmir_eeri_2nd_report.pdf

30) "Nicobar Islands"
http://en.wikipedia.org/wiki/Nicobar_Islands

31) *Quakes, Eruptions and Other Geologic Cataclysms* (*Revealing the Earth's Hazards*, Rev. Ed.) by Jon Erickson (Checkmark Books, NYC, 2001), p. 36.

32) "2004 Indian Ocean earthquake and tsunami"
http://en.wikipedia.org/wiki/2004_Indian_Ocean_earthquake_and_tsunami

33) *The New Catastrophism* by Derek Ager (Cambridge University Press, 1993), p. 195, emphasis added.

34) *Quakes, Eruptions and Other Geologic Cataclysms* (*Revealing the Earth's Hazards*, Rev. Ed.) by Jon Erickson (Checkmark Books, NYC, 2001), p. 3.

35) Ibid., p. 19.

36) "Mid-Cretaceous seafloor spreading pulse: Fact or fiction?" by M. Seton, C. Gaina and R.D. Müller, *Geology*, August 2009; v. 37; no. 8; p. 687–690, p. 687.
http://www.earthbyte.org/people/dietmar/Pdf/Seton_etal_midK_spreading_pulse_Geology09.pdf

37) *Exploring Our Living Planet* by Robert D. Ballard (National Geographic, Wash. D.C., 1983), p. 189.

38) "Fusion by earthquake fault friction: Stick or slip?" by Yuri Fialko and Yakov Khazan, *Journal Geophysical Research*, Vol. 110, 2005, p. 5.
http://igppweb.ucsd.edu/~fialko/Assets/PDF/fialkoJGR05a.pdf

39) "Effect of depth-dependent viscosity on the planform of mantle convection" by Hans-Peter Bunge, Mark Richards and John Baumgardner, *Nature*, Feb. 1, 1996, Vol. 379, pp. 436-438, p. 436, emphasis added.

40) "Evidence suggesting extremely rapid field variation during a geomagnetic reversal" by R.S. Coe and M. Prévot, *Earth and Planetary Science Letters,* 92:292–298 (1989).

41) "New evidence for extraordinary rapid change of the geomagnetic field during a reversal" by R.S. Coe, M. Prévot and P. Camps, *Nature,* 374:687–692 (1995).

42) "Geomagnetic reversals from impacts on the Earth" by Richard Muller and Donald Morris, *Geophysical Research Letters*, Volume 13, Issue 11, pp. 1177–1180, November 1986, p. 1177.
http://onlinelibrary.wiley.com/doi/10.1029/GL013i011p01177/abstract

43) "Geomagnetic reversals?" by Peter Warlow, *Journal of Physics A: Mathematical and General*, 1978, Vol. 11: 2107
http://iopscience.iop.org/0305-4470/11/10/026

44) *Doomsday: the Science of Catastrophe* by Fred Warshofsky (Reader's Digest Press, NYC, 1977), p. 163, emphasis added.

45) "Continental drift: how the knowledge came about and evolved"
http://pratclif.com/geology/contdrift/frame.htm

46) "Antonio Snider-Pellegrini"
http://en.wikipedia.org/wiki/Antonio_Snider-Pellegrini

47) "Integrated 'plume winter' scenario for the double-phased extinction

during the Paleozoic–Mesozoic transition" by Yukio Isozaki, *Journal of Asian Earth Sciences*, Nov. 2009, Vol. 36, Issue 6, pp. 459-480.

48) *Catastrophism* by Richard Huggett (Verso, London, 1997), p. 127.

Chapter 7
Eating Trilobites

Masquerading as a man with a reason
My charade is the event of the season
And if I claim to be a wise man,
Well, it surely means that I don't know
 - Kansas

Students need to be made aware of the human tendency to self-deception in order to avoid the cognitive error of confirmation bias.
 - William D. Stansfield, Cal Poly - San Luis Obispo (2012)[1]

a.f. eol.org

Did trilobites and mammals coexist? Did man and dinosaurs live at the same time? On the show *Fear Factor*, contestants ate worms and bugs. Could it be that early humanity ate trilobites? *Serolis trilobitoides* (above) resembles a trilobite as do other serolid isopods. If Young Earth Science (YES) is real and this planet is thousands (NOT billions) of years old, then people eating trilobites is not an issue.

Out of Place – Out of Time

In *Forbidden Archeology: The Hidden History of the Human Race*, Michael Cremo and Richard L. Thompson (Ph.D. Math, Cornell University) provide evidence for human existence in rocks dated millions of years old.

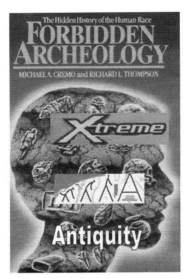

a.f. amazon.com

Recently, a resident of Vladivostok (Russia) found what appears to be a small metal piece from a gear found in a lump of coal! This coal is from the Chernogorodskiy mines of Khakasia region (300M years old). The gear piece is 98 percent Aluminum and 2 percent Magnesium.[2] Thus, we have evidence of humans living 300M years ago by mainstream dating!!

a.f. voiceofrussia.com

Michael Cremo reports of an amazing find:

> In 1830, <u>letterlike shapes</u> were discovered within a solid block of marble from a quarry near Norristown, Pennsylvania, about 12 miles northwest of Philadelphia. The marble block was taken from a depth of 60-70 feet. This was reported in the *American Journal of Science and Arts* (v. 19, p. 361) in 1831. The marble in the quarries around Norristown is Cambro-Ordovician (Stone 1932, p. 225), or <u>about 500-600 million years old</u>.[3]

Did men and women roam this planet during the Cambrian? And, if so, did they eat trilobites?

The April 2, 1897 edition of the *Daily News* of Omaha, Nebraska, carried an article titled "Carved Stone Buried in a Mine," which described an object from a mine near Webster City, Iowa. The article stated:

> While mining coal today in the Lehigh coal mine, at a depth of 130 feet, one of the miners came upon a piece of rock which puzzles him and he was unable to account for its presence at the bottom of the coal mine. The stone is of a dark grey color and about two feet long, one foot wide and four inches in thickness. Over the surface of the stone, which is very hard, lines are drawn at angles forming perfect diamonds. The center of each diamond is a fairly good face of an old man.[4]

The Lehigh coal is probably from the Carboniferous. Could this human artifact really be hundreds of millions of years old?

Living Fossils

Often organisms appear to have gone extinct based on their fossil record. Yet we see them alive today. Why during many millions of years of existence did they escape fossilization? In 1994, a living Wollemi pine was discovered in Australia.

Wollemi pines were thought to have gone extinct at the end of the Cretaceous (65M).[5, 6] If these pines survived from the "Dinosaur Age" until now, maybe dinos have as well.

a.f. zoologischemededelingen.nl

A living fossil of the insect xylastodorine subfamily was discovered in New Caledonia in 1978. The find, *Proxylastodoris kuscheli*, was previously known only from Baltic amber of 40M years ago.[7] So, this represents a range extension of 40M years.

Francis Schaeffer Hall

The coelacanth, thought to have gone extinct 65M years ago, is found off the coasts of South Africa, Kenya, Madagascar and Indonesia.[8] With such a wide distribution, how could they have escaped fossilization for 65M years?! Thus, the lack of a fossil record does not imply that the organism, including man, did not exist. Are dinosaurs walking among us?

a.f. amazon.com

According to Katherine Courage, contributing editor for *Scientific American*, there may be a freshwater octopus in Oklahoma's lakes:

> This unlikely animal, people have explained, might be a rare living fossil, left over from the time (tens of millions of years ago) when this part of the country was, indeed, a shallow sea - and a perfect octopus habitat. Over the millennia, this particular line of octopuses has adapted to freshwater, these proponents suggest.[9]

Which is more likely, that the octopus population survived for millions of years or that these are remnants from a global ocean just thousands of years ago (a SERGA, see Ch. 5)? Could evolutionary dating be completely wrong?

Dinos NOW!

The Vale of Kashmir lies in the high mountainous region between India and Pakistan. The Hindu *Nilamata Purana* (probably composed between AD 550 and AD 700) describes the Valley of Kashmir as once containing an enormous lake. The *Purana* states that a water serpent named Jalodbhava ("Born of Water") inhabited this lake, depopulating the villages near the water because he was an "eater of human flesh."[10] Could this creature have been a large reptile such as a mosasaur? The longest mosasaur reached 57 feet![11]

a.f. deviantart.net

In 2007, in a survey of over 1,000 Canadians, 42% agree that dinosaurs and human beings co-existed on earth![12]   Has *Scientific American* ever supported the thesis that dinos have survived into modern times? In 1922, that publication reported the account of an American gold prospector who saw what appeared to be a plesiosaur in a lake in the Patagonian region of Argentina.[13]

a.f. nbcnews.com

Some Chinese have been using dino fossils in recent years, calling them "dragon bones," in traditional medicines. Dong Zhiming, professor at the Institute of Vertebrate Paleontology and Paleoanthropology of the Chinese Academy of Sciences, explains, "They had believed that the 'dragon bones' were from the dragons flying in the sky."[14]   There is a superb website on cryptozoology (http://www.cryptozoology.com/).

a.f. smithsonianmag.com

At the Ta Prohm temple in Cambodia there is a certain representation of a stegosaur. This is a strong indication that man and dinosaur have coexisted in the not too distant past.[15]

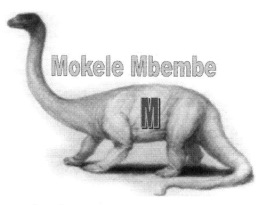

a.f. qsl.net

Numerous eyewitness accounts support the existence of a sauropod, Mokele-Mbembe, in the Congo. In 1981, American engineer Herman Regusters went to Lake Tele and saw a sauropod. The Regusters expedition returned with droppings and footprint casts. He made an audio recording of a low roar thought to be from Mokele-Mbembe. The recording was analyzed, but did not match any known animal.[16]

The latest edition of Rory Nugent's book *Drums Along the Congo: On the Trail of Mokele-Mbembe, the Last Living Dinosaur* appeared in 2013. He claims to have seen the beast.[17]

imdb.com

In the 1985 film *Baby: Secret of the Lost Legend*, a paleontologist and her husband discover sauropods in Africa (mother and child) and protect them from poachers.[18] Issy or Issie (イッシー—Isshī) is a Japanese monster spotted in Lake Ikeda, near Kyushu Island in Japan. It is described as being saurian in appearance.[19]

J. Talboys Wheeler describes a discovery of Herodotus (d. 425 BC): "In Arabia, near Buto, Herodotus saw a heap of bones and spines of winged serpents, said to be killed by an ibis while trying to enter Egypt. ... The winged serpent is like a water-snake with bat's wings."[20] Sounds like a flying reptile sighting, doesn't it?

a.f. *Ancient Chinese Inventions*

Mike Edwards, writing in *National Geographic*, describes Marco Polo's journey to the East,

> Around the city of Dali, for example, was there really a giant snake with legs and a mouth "so large that it would well swallow a man"? … [Shi Lizhou said] "According to legend there was a huge snake with legs here – a snake, not a crocodile – that ate people." Marco also claimed that people here dined on raw flesh … Any doubts I may have had about that evaporated in a village near Dali where I watched a hundred persons sit down to feast on raw pork.[21]

We believe Marco Polo regarding raw meat, why don't we accept his claim to have seen an apparent dinosaur? *Man vs. Wild* star Bear Grylls occasionally eats raw meat!

Even Charles Lyell, the famed uniformitarian, admitted that pterodactyls may fly the modern skies:

> Then might those genera of animals return, of which the memorials are preserved in the ancient rocks of our continents. The huge iguanodon might reappear in the woods, and the ichthyosaur in the sea, while the pterodactyl might flit again through umbrageous groves of tree ferns.[22]

In 1976, three teachers saw a flying creature with a wingspan of 15-20 feet outside of San Antonio. They later identified the flying reptile as a pteranodon.[23]

a.f. *Mystery in Acambaro*

In 1944, over 30K ancient objects were discovered in Acambaro, Mexico made from ceramic and stone. Some of these resemble dinos and flying reptiles. Charles Hapgood, advocate of the pole shift theory, visited Acambaro in 1955 and 1968 and became convinced that the figurines were authentic. A number of dating methods placed the figurines to around 2000 BC.[24]

Range Extensions

A stratigraphic range extension is when a fossil is found in older or younger rocks compared to the usual time span an organism's remains are found. Could dinos still exist today? Could man have lived during the Cambrian?

A mouse-sized primate fossil has recently been discovered in China and dated at 55M years.[25] So, the last dino and the first primate were not that far apart! A team of Canadian scientists reported in *Palaeontology* about fossil horseshoe crabs, dated at 445M years old, which is about 100 million years older than the prior estimate![26] Could dino fossils exhibit a similar stratigraphic range extension and be found in recent strata? A 70M year old duck-like bird fossil has been discovered. We normally don't think of dinos and ducks hanging out together.[27] So is it so unusual to posit that man and trilobites coexisted?

Haramiyid mammals date to between 220M to 200M years ago.[28] The most recent trilobite existed 245M years ago.[29] So the gap may be as short as 25M years! If the first mammal and the last trilobite may have coexisted, is it such a stretch to think that man and trilobites lived at the same time? A pterosaur was discovered in 1973 by Mario Pandolfi in Italy. It was the oldest pterosaur then known (203M – 210M). Prior references had placed the pterosaurs in the Cretaceous.[30] So this is a major stratigraphic range extension of dozens of millions of years deeper into the past! What's next – man with trilobites?

Michael Engel and David Grimaldi describe an amazing insect discovery:

> Only a few fossils provide insight into the earliest stages of insect evolution, and among them are specimens in chert from Rhynie, Scotland's Old Red Sandstone (Pragian; about 396 - 407 million years ago) … For <u>true insects</u> (Ectognatha), the oldest records are two apparent wingless insects from later in the <u>Devonian</u> period of North America. Here we show, however, that a fragmentary fossil from Rhynie, *Rhyniognathahirsti*, is … the earliest true insect …*Rhyniognatha* indicates that <u>insects originated in the Silurian</u> period and were members of some of the earliest terrestrial faunas.[31]

This time gap represents approximately 300M years!! So the absence of fossils for a given era does not mean an organism did not exist at that time. Could early man have eaten trilobites?

George Stanley Jr. reports a significant range extension:

> Tiny cap-shaped fossils from the Upper Triassic rocks of Vancouver Island (Wrangellia) are identified as a calcareous sponge *Nucha? vancouverensis*... The genus *Nucha*... was first reported from Middle Cambrian rocks of Australia. The Triassic example constitutes the first occurrence outside the Cambrian system and the continent of Australia.[32]

This represents a range extension of hundreds of millions of years!! Could this evidence lead credence to the co-existence of humans and trilobites?

Robert MacNaughton et al provide another fantastic range extension: "In the Cambrian-Ordovician Nepean Formation near Kingston, Ontario, arthropod-produced trackways extend the range of arthropods on land by as much as 40M years into the past."[33] Chinese paleontologists have discovered fish fossils from the Lower Cambrian! This is 50M years before the current estimate of when fish were previously known (530M).[34]

Larry Heaman (Department of Earth and Atmospheric Sciences at the University of Alberta) and his team found a hadrosaur bone in New Mexico dated at 64.8M. That makes it 700K years after the mass extinction that killed most of the dinos at the KT boundary.[35] Does this uphold the suspicion of many that dinosaurs may roam the world today in some isolated spot?

a.f. YouTube

In a paper presented at the International Congress of History o Science in Mexico City in 2001, Michael Cremo presented evidence of strata in the "wrong" order. The age of the Salt Range Formation in the Salt Range Mountains of Pakistan is in dispute. The Cambrian age of the overlying Purple Sandstone is evidenced by trilobites. Dicotyledonous leaves have been found in the underlying Salt Range Formation which are advanced plants and did not appear until the Cretaceous. There is no evidence of a massive overthrust to explain the reverse order of the rock layers. Fossil wood, Tertiary forams and even an *insect* have been found in the Salt Range Formation! In 1944, B. Sahni gave the presidential address to India's National Academy of Sciences and reported on pollen, wood fragments, and insect parts found in the Salt Range Formation.[36]

Absence of Evidence

The non-marine bivalve *Anthraconauta* in absent in the following zones in the British Upper Carboniferous: Prolifera, Similis-Pulchra and Modiolaris. *Anthraconauta* is only partially present in the Communis and Lenisulcata zones.[37] Does this absence mean that this bivalve vanished off the face of the Earth? If human bones are not found in the Cambrian, does this mean that the Cambrian fauna pre-date the dawn of mankind?

Zones	Abundance		
Prolifera	?	?	?
Tenuis		▓	
Phillipsii	▓	▓	
Similis-Pulchra	?	?	?
Modiolaris	?	?	?
Communis		▓	
	?	?	?
	?	?	?
Lenisulcata		▓	

Abundance of *Anthraconauta*
Upper Carboniferous (UK)
data from *The New Catastrophism*

In 1966, Robert Silverberg stated, "There are no fossil fishes known that closely resembled *Protopterus* or *Lepidosiren*." *Lepidosiren* is the South American lungfish.[38] Because no fossils had been found at that time, did that imply that this variety of lungfish did not exist during the Cretaceous? Likewise, will human fossils someday appear in the Devonian? At one time, millions of buffalo roamed across America – where are the buffalo fossils?

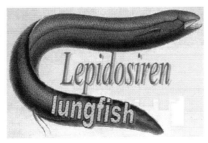

a.f. wikipedia

Does this evidentiary smorgasbord support the view that humans and other advanced animals existed during the Cambrian? Was man eating trilobites? Are there living dinosaurs? Clearly, women eating trilobites is consistent with YES. Given that YES is massively endorsed by the observational evidence, what does the future hold for Young Earth Science? This question is our next topic.

Notes:
1) "Science & the Senses: Perceptions & Deceptions" by William D. Stansfield, *The American Biology Teacher*, Volume 74, Issue 3, March 1, 2012, pp. 145-149, p. 145.
2) "300-million-year-old UFO tooth-wheel found in Russian city of Vladivostok" by Yulia Zamanskaya
http://voiceofrussia.com/2013_01_19/300-million-year-old-UFO-tooth-wheel-found-in-Russian-city-of-Vladivostok/
Accessed 10-16-14.
3) *Human Devolution* by Michael Cremo, (Torchlight Publishing, 2003), use "See Inside" feature, emphasis added.
http://www.amazon.com/Human-Devolution-Alternative-Darwins-Theory/dp/0892133341
4) Ibid.
5) "The Lost Valley of the Wollemi Pine" by Jennifer Frazer
http://blogs.scientificamerican.com/artful-amoeba/2012/06/30/the-lost-valley-of-the-wollemi-pine/
6) "Ancient Wollemi Pines Resurgent" by Stephen McLoughlin and Vivi Vajda, *American Scientist*, Vol. 93, Nov.-Dec. 2005, pp. 540-547, p. 540.
7) "Discovery of a species from New Caledonia (Heteroptera: Thaumastocoridae) and first record of the subfamily from the eastern Hemisphere" by P.H. van Doesburg, G. Cassis and G.B. Monteith, *Zoologische Mededelingen Leiden* 84 (2010), pp. 93-115.
8) "Ancient Swimmers" by Carolyn Butler, *National Geographic*, March 2011, Vol. 219, No. 3, pp. 86-93, p. 88.
9) "Could an Octopus Really Be Terrorizing Oklahoma's Lakes?" by Katherine Courage
http://blogs.scientificamerican.com/octopus-chronicles/2013/12/19/could-an-octopus-really-be-terrorizing-oklahomas-lakes/
10) "Kashmir" by David Mathisen
http://mathisencorollary.blogspot.com/2011/06/kashmir.html
11) "Mosasaur"
http://en.wikipedia.org/wiki/Mosasaur
12) "Do Canadians Believe in Evolution Or Creationism?" by Angus Reid Strategies
http://www.angus-reid.com/wp-content/uploads/archived-pdf/ARS_Evo_Cre.pdf
13) "Is the Argentine Plesiosaurus a Fake or a Scientific Marvel" by Leonard Matters, *Scientific American*, Vol. 127 (1922), p. 21.
14) "Chinese villagers ate dinosaur 'dragon bones'"
http://www.nbcnews.com/id/19606626/ns/world_news-asia_pacific/t/chinese-villagers-ate-dinosaur-dragon-bones/#.Url69n90bAI
15) *Ancient Angkor* by Michael Freeman and Claude Jacques (River Books, Bangkok, 2006), pp. 142, 143.

16) "Mokele-mbembe"

http://en.wikipedia.org/wiki/Mokele-mbembe

17) "Drums Along the Congo: On the Trail of Mokele-Mbembe, the Last Living Dinosaur"

http://www.amazon.com/Drums-Along-Congo-Mokele-Mbembe-ebook/dp/B00E46LYAG

18) "Baby: Secret of the Lost Legend"

http://www.imdb.com/title/tt0088760/

19) "Issie"

http://en.wikipedia.org/wiki/Issie

20) *An Analysis and Summary of Herodotus* by J. Talboys Wheeler (Henry Bohn, London, 1852), p. 64, Book 2:75, 76.

http://books.google.com/books?id=Pv0NAAAAYAAJ&pg

21) "Marco Polo in China (Part II)" by Mike Edwards, *National Geographic*, June 2001, Vol. 199, no. 6, pp. 20-45, p. 33.

22) Quoted in "The Fall and Rise of Catastrophism" by Trevor Palmer (1996), emphasis added,

http://archive.is/7LuiT

23) *Mystery in Acambaro* by Charles Hapgood and David Hatcher Childress (Adventures Unlimited Press, Kempton, IL, 2000), p. 45.

24) Ibid., pp. 13-18.

25) "Tiny Chinese Archicebus fossil is oldest primate yet found" by Jonathan Amos,

http://www.bbc.co.uk/news/science-environment-22770646

26) "Oldest Horseshoe Crab Fossil Found, 445 Million Years Old"

http://www.sciencedaily.com/releases/2008/02/080207135801.htm

27) "Cretaceous duck ruffles feathers"

http://news.bbc.co.uk/2/hi/science/nature/4187287.stm

28) "Ancient Squirrel-Like Creatures Push Back Mammal Evolution" by Charles Choi,

http://www.livescience.com/47774-ancient-squirrels-push-back-mammal-evolution.html

29) "Trilobita: Fossil Record"

http://www.ucmp.berkeley.edu/arthropoda/trilobita/trilobitafr.html

30) "Eudimorphodon"

http://en.wikipedia.org/wiki/Eudimorphodon

31) "New light shed on the oldest insect" by Michael Engel and David Grimaldi, *Nature* 427(6975):627–630, 2004, p. 627, emphasis added.

http://www.nature.com/nature/journal/v427/n6975/abs/nature02291.html

32) "Triassic sponge from Vancouver Island: possible holdover from the Cambrian" by George Stanley Jr., *Canadian Journal of Earth Sciences* 35(9):1037–1043, 1998, p. 1037, emphasis added.

http://www.nrcresearchpress.com/doi/abs/10.1139/e98-049?journalCode=cjes#.Uhge6qx0bAI

33) "First steps on land: Arthropod trackways in Cambrian-Ordovician eolian sandstone, southeastern Ontario, Canada" by Robert MacNaughton,

Jennifer Cole, Robert Dalrymple, Simon Braddy, Derek Briggs and Terrence Lukie, *Geology* 30(5):391–394, 2002, p. 391, emphasis added.

34) "Oldest fossil fish caught"
http://news.bbc.co.uk/2/hi/sci/tech/504776.stm

35) "Dinosaurs Survived Mass Extinction by 700,000 Years, Fossil Find Suggests"
http://www.sciencedaily.com/releases/2011/01/110127141707.htm

36) "Paleobotanical Anomalies Bearing on the Age of the Salt Range Formation of Pakistan" by Michael A. Cremo, Presentation at XXI International Congress of History of Science, Mexico City, July 8-14, 2001
http://www.mcremo.com/saltrange.html

37) *The New Catastrophism* by Derek Ager (Cambridge University Press, 1993), p. 135.

38) *Forgotten by Time* by Robert Silverberg (Thomas Crowell Co., NYC, 1966), p. 65.

Chapter 7
YES, the Future

A genuine leader is not a searcher for consensus, but a molder of consensus.
- Martin Luther King Jr.

All great scientists ignore falsification [~Popper]. … you ignore what others say; you stick to your suspicions and beliefs and you seek to prove what you secretly know in advance to be true. … All scientists are advocates …
- Terence Kealey (Vice-Chancellor, University of Buckingham)[1]

Trust Me – I'm an Expert

On Sunday, October 30, 1938 (Halloween eve) Orson Welles and the Mercury Theater aired "War of the Worlds" by H.G. Wells and created a panic. Perhaps as many as a million radio listeners believed that the Martians actually invaded.[2] One listener said, "I believed the broadcast as soon as I heard the professor from Princeton and the officials in Washington."[3] Are we so quick to trust the experts that we stop applying our critical thinking skills? Have we been deceived by Old Earth Fallacies (OEF's)? We have found no Martians on Mars; likewise, there is a lack of solid evidence for an Old Earth.

The War of the Worlds (2005)

Has Big Science ever been wrong? Antoine Lavoisier, the father of modern chemistry said, "Stones cannot fall from the sky, because there are no stones in the sky!" There are very few meteorites in museums that were found before 1790.[4]

Some OEF conformists think that there is no such thing as a Young Earth. Some exaggerate Young Earth Theory and throw insults at it. However, the clues are there for those willing to see it. Many of the skeptics regarding rocks from the sky were convinced, when on April 26th, 1803 in L'Aigle, France more than two thousand rocks fell from the sky.[5] The Pro-YES (Young Earth Science) evidence is falling all around us. I say, "Let the rocks fall!"

In the November 1999 issue of *National Geographic*, Archaeoraptor was introduced as a dino with feathers. Kudos to *National Geographic* for giving five pages of the October 2000 issue to chronicle the research that proved that Archaeoraptor is a fake. Lewis Simons explains, "To some prominent paleontologists who saw it, though, the little skeleton was a long-sought key to a mystery of evolution. To others among this frequently hirsute and determinedly individualistic fraternity, it was a cheap hoax."[6] Evolutionist are so quick to accept evidence that supports Darwin because of their PreSuppositions. Shoulgd we avoid the fuzzy data presented to push OEF's and shelve our bias against YES and fairly consider the facts that favor YES.

Radiocarbon in Coal

If you have read this whole book and you are still unconvinced of YES, we invite you to go to your nearest coal mine and have a sample dated. We predict that the radiocarbon age will be less than 50K years. So, coal supports YES. We read on radiocarbon.com the following:

> Radiocarbon activity of materials in the background is also determined to remove its contribution from results obtained during a sample analysis. Background radiocarbon activity is measured, and the values obtained are deducted from the sample's radiocarbon dating results. Background samples analyzed are usually geological in origin of infinite age such as coal, lignite, and limestone.[7]

a.f. teeic.anl.gov

If we take the background radiocarbon at face value, we conclude that coal is of **recent origin** and not millions of years old.

Who's Dating Your Planet?

Other methods can be applied to Cambrian rocks to verify their recent origin. Optically Stimulated Luminescence Dating can date rocks that are considered to be several hundred thousand years.[8] If Ordovician rocks are in actual fact only thousands of years old, then this method can be used to falsify the current dating of the scientific elite.

The electron spin resonance dating method can be used to date stalagmitic calcite, shells, bones and teeth. This method has indicated that modern humans have existed in Israel and Africa for more than 100K years. Could this technique be used to show that fossils, allegedly millions of years old, are really much younger?[9]

Volcanic signatures based on the ratios of rare earth elements may be used to correlate strata. The relative amount of Cs, Rb, Ba, Th, U, Nb, K, La and so on, form a unique pattern for specific lava flows.[10] Could this technique be used to verify that most of the rock record was part of a SERGA (**S**ingular **E**poch of **R**apid **G**eologic **A**ctivity)?

YES, the Future of Research

Who will support YES research? What is the position of the National Science Foundation (NSF) towards YES? David B. Campbell a program director with the NSF provides an answer:

> In response to your inquiry for research funding for Young Earth Science, the Geoscience directorate at the NSF supports a variety of research in the earth and atmospheric sciences. NSF would support research that investigates the age of the earth in the form of a research proposal that describes hypotheses and rigorous methods to test those hypotheses. All proposals are peer reviewed by scientific experts from across the U.S., using the National Science Board merit

review criteria regarding intellectual merit and broader impacts.[11]

It is worth noting that the NSF links to an anti-YES blog.[12] The NSF clearly states that you are leaving their site and that they do not necessarily endorse the views of off-site blogs; however, this diatribe is revealing: "We can't imagine vast stretches of time, such as a few million years in the past when our ancestral hominids roamed the earth. It's easier and more comforting to reject science and think in a young earth timeframe..."[13] If the NSF is slow regarding YES research, will other organizations stand up for the challenge and investigate the facts that favor YES? Here are a few possibilities for research grants:

3M	BP	Exxon	GE
BASF	Dupont	Alon	Genentech
Microsoft	Apple	Monsanto	IBM
Dow	DuPont	Chevron	Celera

Leave your OEF research dungeon and join the growing cohort of Young Earth Scientists.

Tree Rings have the Ring of Truth

Fossil logs with growth rings were analyzed from the Upper Permian and Middle Triassic of the central Transantarctic Mountains. Tree rings in both the Permian and Triassic woods show similar ring pattern with mostly earlywood, indicating a temperate climate.[14] So the Permian and Triassic had similar weather. Could this study be carried further by correlating these trees using standard dendrochronology methods? This would substantiate the claim that the Upper Permian and Middle Triassic are **not** separated by millions of years.

Global Catastrophe Soon?

Will another planet-wide disaster strike the earth soon? Isaac Asimov wrote a book, *A Choice of Catastrophes*, where he described various possible endings. There are a number of Near Earth Objects we should be concerned about. If the Yellowstone supervolcano erupted, it would bring about great destruction.   Will a global cataclysm affect this planet sometime in the next one hundred years? Time will tell.

Charles Hadlock has written *Six Sources of Collapse* which surveys potential natural and man-made disasters as well as extreme weather events.

There are numerous existential risks this planet faces. There are tons of films with an "end of the world" theme, but could it happen in real life? Global catastrophe is a genuine hazard. We have provided clues that there was a Global cataclysm just a few thousand years ago – will another arrive before long? According to the BBC's Laurence Peter,

> …the chances of a NEO [Near Earth Object] smashing into Earth are very slim, it would only take one to cause a global catastrophe and perhaps wipe out mankind. Even one smaller than 140m in diameter could cause a continental disaster - a giant fireball or a tsunami.[15]

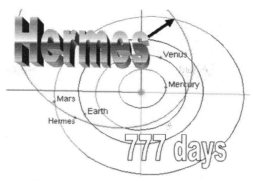

a.f. nasa.gov

Hermes is a double asteroid (1937 UB), two space rocks orbiting one another, each about 400 meters across!! In 1937, 1942, 1954, 1974 and 1986, Hermes came close to Earth. Asteroid Hermes travels to the inner solar system every 777 days.[16] Each piece of Hermes would create epic devastation if it hit the Earth. Writing in 2012, Elisabetta Pierazzo and Natalia Artemieva of the Planetary Science Institute of Arizona warned, "Modern civilization is vulnerable to even relatively small impacts, which may occur in the near future, that is, tens to hundreds of years."[17] So there is a decent chance that an asteroid or comet could completely destroy civilized life as we know it within your lifetime!!

Meteors about the size of the one that exploded over Chelyabinsk in 2013 may be up to seven times more likely to hit the planet than mainstream's prior estimates. In 1908 a giant blast happened in Siberia and in 1963 an airborne explosion hit near South Africa. It was once thought that these types of violent events came once every 8K years, yet they all occurred in a 105-year time span.[18] Could this finding also be a clue that the dating methods of Big Science are fundamentally flawed? That's a 98.7% reduction in time. Should the standard geologic time scale be reduced by a similar amount?

a.f. wikipedia

Conform and Control

The media manipulators of Big Science dominate the cultural voice and spread OEF's far and wide. Help deter this mass deception by distributing copies of this book to your friends and family members as well as scientists, medical professionals, media personnel, bloggers, YouTubers, engineers and educators you may know.

Say "NO" to H8RS!

Anti-YES messages are bullying our children. Stand up and defend your position. Open discussion is better than anonymous insults. If you google "young earth" and "flat earth" together you get over 23K hits. This vile oppression must stop! Phil Plait writing at DiscoverMagazine.com says that YES advocates are promoting "antiscientific nonsense."[19] Berkley's pro-evolution "Understanding Science" website disdains Young Earth Theory and calls it a "non-scientific viewpoint."[20]

This uncouth bullying is a disgrace to those who claim to be tolerant, pro-discussion and against compulsion. Deep Time bullies should turn their "Ha Ha" into "Aha!" as they feel the power of Young Earth evidence.

An article in *Global and Planetary Change* (Elsevier) stated with disdain, "By the 1800s ... most eminent savants acknowledged that sufficient natural evidence was in place to discredit the young-earth theory as a factual absurdity ..."[21] Does such caustic rhetoric constitute bullying? David Oldroyd of the School of History and Philosophy of Science at the University of New South Wales, calls ambassadors of Young Earth Theory "fools."[22] Writing in the journal *Evolution*, Sehoya Cotner et al confirm the trend of slashing out against Young Earth Theory scholars: "...adherence to young-Earth [theory] requires a refutation not only of modern biology, but also geology, paleontology, and physics, such convictions may serve as a proxy for scientific ignorance in general."[23]

Let's all reject anti-YES bullying and second the thoughts of Geology Professor F. N. Earll on Hapgood's pole shift theory and apply them to Young Earth Science (YES):

> Let us not bury this idea prematurely through prejudice, as so many valuable ideas of the past have been buried, only to be sheepishly exhumed in later years. If it is an unworthy thing let it be properly destroyed; if not, let it receive the nourishment that it deserves.[24]

Despite this bullying, YES is a strong force in today's world. Melvin Bragg (Chancellor of the University of Leeds) hosted a BBC radio program on the Age of the earth.[25]   Several comments supported YES. There were apparently seven who favored YES and seven who agreed with the Old Earth view. Is the BBC pro-YES? Maybe not, but at least they gave the YES advocates an opportunity to argue their case.[26]

Some scholars actually show a respectful attitude toward YES. R. F. Diffendal Jr. in his article "Earth in Four Dimensions: Development of the Ideas of Geologic Time and History" in *Nebraska History* presents the controversy fairly:

> To some extent the arguments about a short Earth history of a few thousand years versus a long history of billions of years are still going on today. Most natural scientists support the long history. However, some people including Richard Milton [Mensan, agnostic, science journalist] … support the idea of a short one. Many also think and write about time and our place in it (e.g. G. G. Simpson in *The Dechronization of Sam Magruder*).[27]

Bold researchers are resisting the peer pressure of OEF professors and seeking out the truth about YES.

YES Advocates of the World Unite!

Let's take a new path and share the overwhelming evidence for YES and leave no rock behind! Imagine the world of the future where eager kids learn the truth about YES and reject OEF's. OBEY the urge to say "YES!" Free Geology from the shackles of the old earth paradigm. Reject the epistemic closure of OEF's and join the excitement of the YES outreach adventure! Escape from the Deep Time cult and expose OEF's. We predict that the grandiose evolutionary timescale will fall. Are those unwilling to hear the evidence for YES, the genuine "owners of a lonely heart?"

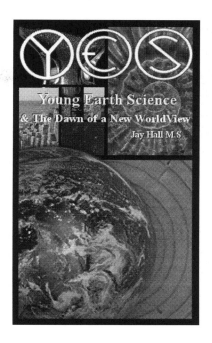

How can you help promote YES? One way is to assist the adoption of YES as an acronym for Young Earth Science. Please send all uses of YES used in this way to,

Merriam-Webster
POB 281
Springfield MA 01102

Follow the path of parsimony and investigate YES. Intelligent Catastrophism (IC) is the least complex explanation for earth history. Join the young earth detectives and solve the case of "Earth's Missing Birthday." A reasonable man or woman should realize that much geologic evidence supports YES. Rapid Plate Movements (RPM) may seem like a radical hypothesis, but so was continental drift. Will YES someday achieve full-scale acceptance and be considered firmly established by science? Our hope is YES!

John Stuart Mill, in *On Liberty*, significantly stated that, "… the majority of eminent men of every past generation held many opinions now known to be erroneous …"[28] Mill opposed an "atmosphere of mental slavery" and advocated open dialog:

Where there is a tacit convention that principles are <u>not</u> <u>to be disputed</u>; where the <u>discussion of the greatest</u> <u>questions which can occupy humanity is considered to</u> <u>be closed</u>, we cannot hope to find that generally high scale of mental activity which has made some periods of history so remarkable.[29]

Mill further suggests that we should investigate both sides of a scientific controversy:

<u>Even in natural philosophy, there is always some other</u> <u>explanation possible of the same facts</u>; some geocentric theory instead of heliocentric, some phlogiston instead of oxygen; and <u>it has to be shown why that other theory</u> <u>cannot be the true one</u>: and until this is shown, and until we know how it is shown, <u>we do not understand the</u> <u>grounds of our opinion</u>.[30]

In Conclusion, we concur firmly with a report from SINTEF (independent research organization in Norway) by Emil Røyrvik:

In open societies where both scientists and the general public are equipped with critical skills and the tools of inquiry, not least enabled by the information revolution provided through the Internet, the ethos of science as open, questioning, critical and anti-dogmatic should and can be defended also by the public at large. Efforts to make people bow uncritically to the authority of a dogmatic representation of Science, seems largely to produce ridicule, opposition and inaction, and ultimately undermines the legitimacy and role of both science and politics in open democracies.[31]

Notes:
1) "Dr. Terence Kealey Presentation"
http://www.youtube.com/watch?v=CyOB1KDzfYk
2) "Welles scares nation"

http://www.history.com/this-day-in-history/welles-scares-nation

3) *The War of the Worlds* by H.G. Wells (Sourcebooks, Naperville, IL, 2005), p. 14.

4) *Alternative Science: Challenging the Myths of the Scientific Establishment* by Richard Milton (Park Street Press, Rochester, Vermont, 1996), p. 3.

5) Ibid., p. 3

6) "Archaeoraptor Fossil Trail" by Lewis Simons, *National Geographic*, October 2000, pp. 128-132, p. 128.

7) "Radiocarbon Dating: An Introduction," emphasis added.
http://www.radiocarbon.com/about-carbon-dating.htm

8) "Optically Stimulated Luminescence Dating of Sediments over the Past 200,000 Years" by Edward Rhodes, *Annual Review of Earth and Planetary Sciences*, Vol. 39 (2011): 461-488, p. 461.
http://www.annualreviews.org/doi/abs/10.1146/annurev-earth-040610-133425

9) "Electron Spin Resonance (ESR) Dating of the Origin of Modern Man" Henry Schwarcz and Rainer Grun, *Philosophical Transactions: Biological Sciences*, Vol. 337, No. 1280 (Aug. 29, 1992), pp. 145-148.
http://www.jstor.org/discover/10.2307/57101?uid=3739920

10) "The Two Main Types of Fingerprint" by B.M. Gunn
http://www.geokem.com/finger.html

11) Personal email from David B. Campbell, 9-9-14.

12) NSF link to anti-YES blog,
http://webcache.googleusercontent.com/search?q=cache:RXxyU7lCX0sJ:www.nsf.gov/news/ipy/DesktopModules/Articles/LeaveReminder879c.html%3FmyURL%3Dhttp://icelabyrinth.blogspot.com/+&cd=15&hl=en&ct=clnk&gl=us

13) "Ice Labyrinth blog" ed. by Sam Bowser, emphasis added.
http://icelabyrinth.blogspot.com/

14) "Tree growth at polar latitudes based on fossil tree ring analysis" by Edith Taylor and Patricia Ryberg, *Palaeogeography, Palaeoclimatology, Palaeoecology,* Nov. 2007, Vol. 255, p246-264.
http://www.sciencedirect.com/science/article/pii/S003101820700332X

15) "Hunt for space rocks intensifies" by Laurence Peter
http://news.bbc.co.uk/2/hi/science/nature/7824002.stm

16) "The Curious Tale of Asteroid Hermes"
http://science1.nasa.gov/science-news/science-at-nasa/2003/31oct_hermes/

17) "Local and Global Environmental Effects of Impacts on Earth" by Elisabetta Pierazzo and Natalia Artemieva, *Elements*, Feb. 2012, Vol. 8, pp. 55-60, p. 55.

18) "Russian fireball shows meteor risk may be bigger" by Seth Borenstein, Nov. 7, 2013,
http://www.boston.com/news/science/2013/11/07/russian-fireball-shows-meteor-risk-may-bigger/aZWm8NBVFX5jyvMdUP6KnI/story.html

19) "The US Congress Anti-Science Committee" by Phil Plait

http://blogs.discovermagazine.com/badastronomy/2012/10/06/the-us-congress-anti-science-committee/#.Uob7JH90bAI

20) "What controversy: Is a controversy misrepresented or blown out of proportion?"
http://undsci.berkeley.edu/article/0_0_0/sciencetoolkit_06

21) "Geoscience meets the four horsemen?: Tracking the rise of neocatastrophism" by Nick Marrinera, Christophe Morhangea, Stefan Skrimshireb, *Global and Planetary Change*, Vol. 74, Issue 1, Oct. 2010, pp. 43-48, p. 46, emphasis added.

22) "The Chronologers' Quest" [book review] by David Oldroyd, *Episodes*, Mar. 2007, Vol. 30, no. 1, pp. 64-65, p. 65.

23) "Is The Age Of The Earth One Of Our 'Sorest Troubles?' Students' Perceptions About Deep Time Affect Their Acceptance Of Evolutionary Theory" by Sehoya Cotner, D. Christopher Brooks and Randy Moore, *Evolution*, Vol. 64, no. 3, pp. 858-864, p. 863, emphasis added.
http://www.cbs.umn.edu/sites/default/files/public/downloads/age-of-earth-evo-march2010.pdf

24) *The Path of the Pole* by Charles Hapgood (Chilton Book Co., Philadelphia, 1970), Rev. Ed. of *Earth's Shifting Crust* (1958), p. ix.

25) "In Our Time - Ageing the Earth"
http://www.bbc.co.uk/programmes/p005493g

26) "In Our Time – Debate"
http://www.bbc.co.uk/radio4/history/inourtime/inourtime_comments_ageing_earth.shtml

27) "Earth in Four Dimensions: Development of the Ideas of Geologic Time and History" by R. F. Diffendal Jr., *Nebraska History* 80 (1999): 95-104, p. 103.
http://www.nebraskahistory.org/publish/publicat/history/full-text/NH1999GeologicTime.pdf

28) *Utilitarianism and On Liberty* by John Stuart Mill, ed. by Mary Warnock (Blackwell Pub., Malden, MA, 2003), p. 102.

29) Ibid., p. 113, emphasis added.

30) Ibid., p. 115, emphasis added.

31) "Consensus and Controversy – The Debate on Man Made Global Warming" by Emil Røyrvik (SINTEF, Trondheim, Norway, 2013), Report A24071, p. 68.
http://www.klimarealistene.com/web-content/Report%20Consensus%20and%20Controversy%2010.0%20.pdf

About the Author

Jay Hall is Assistant Mathematics Professor at Howard College in Big Spring, Texas. He has a Master of Science degree in Mathematics from the University of Oklahoma. Hall has 53 credit hours of Science courses in various disciplines. He has taught at the High School, Technical School and Community College levels. He also has experience in the actuarial field for a number of insurance and consulting organizations. Hall has previously published the Math textbook *Calculus is Easy* (http://www.amazon.com/Calculus-Easy-Jay-Hall-ebook/dp/B00B3YWYNI) and has a paper on MathWorld ("One-Seventh Ellipse," http://mathworld.wolfram.com/One-SeventhEllipse.html). He is also a member of the Choctaw Nation of Oklahoma. You may contact Jay Hall at YoungEarthScience@yahoo.com

Graphics Sources

Note: book covers are from amazon.com

p. 2, Mar. 17, 1952,
http://topics.time.com/mortimer-adler/covers/

p. 3
http://i.telegraph.co.uk/multimedia/archive/01294/down-house2_1294268c.jpg

p. 4
http://www.sfmconsulting.org/images/slideshow/03.jpg

p. 7
http://en.wikipedia.org/wiki/Novum_Organum

p. 8
http://en.wikipedia.org/wiki/Rod_serling
http://en.wikipedia.org/wiki/Vincent_Price

p. 17
http://pubs.usgs.gov/fs/2007/3015/

p. 18
http://researchnews.osu.edu/archive/foscolor.htm
http://www.smithsonianmag.com/science-nature/dinosaur.html

p. 21
http://nrmsc.usgs.gov/Bristlecone
http://3dparks.wr.usgs.gov/grba/html/grba0052.html

p. 22
http://en.wikipedia.org/wiki/Neanderthal

p. 24
http://en.wikipedia.org/wiki/Ra_%281972_film%29

p. 26
Maps of the Ancient Sea Kings by Charles Hapgood
(Adventures Unlimited Press, Kempton, IL, 1996), orig. ed.
1966, p. 81.
http://libweb5.princeton.edu/visual_materials/maps/websites/pacific/cook2/cook2.html
http://en.wikipedia.org/wiki/West_Antarctic_Ice_Sheet

p. 27
http://commons.wikimedia.org/wiki/File:Zeno_1558.png
Maps of the Ancient Sea Kings by Charles Hapgood
(Adventures Unlimited Press, Kempton, IL, 1996), orig. ed.
1966, p. 176.

p. 29
http://www.uscpfa.org/uscr.html
http://www.uscpfa.org/USCR/USCPFA%202011%20Summer.pdf

p. 31
http://en.wikipedia.org/wiki/Awan_dynasty
http://en.wikipedia.org/wiki/Gilgamesh

p. 32
http://www2.econ.iastate.edu/classes/econ355/choi/bab.htm
http://en.wikipedia.org/wiki/Sumerian_language

p. 36
http://en.wikipedia.org/wiki/Roche_limit

p. 37
http://explorers.neaq.org/

p. 39
http://www.hnhs.org/view_meeting.php?id=110
http://www.gettyimages.com/detail/news-photo/william-thomson-lord-kelvin-scottish-mathematician-and-news-photo/113443138

p. 41
http://en.wikipedia.org/wiki/Weald_Basin

p. 51
"A History of Geologic Time" by Henry Faul, *American Scientist*, Vol. 66 (1978), p. 164.
http://faculty.jsd.claremont.edu/dmcfarlane/bio145mcfarlane/PDFs/Faul_History%20of%20Geological%20Time.pdf

p. 53
The Earth's Age And Geochronology by Derek York and Ronald Farquhar (Pergamon Press, Oxford, 1972), p. 46, 85.

p. 55
The Earth's Age And Geochronology by Derek York and Ronald Farquhar (Pergamon Press, Oxford, 1972), p. 110, "Geologic Time Scale Bookmark"
http://pubs.usgs.gov/gip/141/

p. 56
http://www.journals.elsevier.com/earth-and-planetary-science-letters

p. 60
http://www.sciencemag.org/content/282/5395/1840

p. 66
http://en.wikipedia.org/wiki/Devils_Postpile_National_Monument

p. 83
http://en.wikipedia.org/wiki/Ben_Carson

p. 86
http://www.theguardian.com/science/2009/jan/21/charles-darwin-evolution-species-tree-life

p. 87

http://www.landesbioscience.com/books/iu/id/673/?nocache=737851460

p. 94

http://3dparks.wr.usgs.gov/grandcoulee/html/gc699.htm
http://www.pbs.org/wgbh/nova/megaflood/

p. 95

http://en.wikipedia.org/wiki/Ren%C3%A9_Thom

p. 96

http://www.channer.tv/TV_Schedule.htm,%2008-29-11.htm%20NEW%20OK%20%20USE.htm

p. 99

http://meteorite.org/impact-structures.shtml

p. 104

http://en.wikipedia.org/wiki/Surtsey

p. 105

www.geolsoc.og.uk/supereruptions

p. 106

http://video.foxnews.com/v/2695153025001/deadly-earthquake-in-pakistan-creates-new-island

p. 108

John Douglass made a model to show that lake overflow can form a canyon rapidly.
http://www.youtube.com/watch?v=lXHKeJ7VskQ
How the Earth Was Made (History Channel),
http://www.history.com/shows/how-the-earth-was-made/videos

p. 109

http://www2.pvc.maricopa.edu/~douglass/v_trips/grand_canyon/introduction_files/Part3.html

p. 116
http://mathworld.wolfram.com/HyperspherePacking.html

p. 118
http://news.stanford.edu/news/2010/november/van-andel-obit-110810.html

p. 122
http://www.bbc.co.uk/programmes/p005493g

p. 123
Open University presentation on geological time,
http://www.youtube.com/watch?v=fPsklwFdjDY

p. 124
http://academic.emporia.edu/aberjame/map/design.htm

p. 126
http://palaeo.gly.bris.ac.uk/palaeofiles/lagerstatten/

p. 128
http://en.wikipedia.org/wiki/Polystrate_fossil
http://geology.com/publications/lyell/ch23.shtml

p. 129
A Text-book of Geology – Historical Geology (Part 2) by
Charles Schuchert (John Wiley & Sons, NYC, 1915), p. 785
http://books.google.com/books?id=A07kAAAAMAAJ

p. 130
http://en.wikipedia.org/wiki/Turbidite

p. 136
http://nsidc.org/arcticseaicenews/

p. 137
http://p38assn.org/glaciergirl/images/GPR-measurement-concept.jpg

p. 138
http://en.wikipedia.org/wiki/Muir_Glacier

p. 139
http://en.wikipedia.org/wiki/Ice_Age_(2002_film)

p. 140
http://en.wikipedia.org/wiki/Seven_Devils_Mountains

p. 148
The Science of Evolution by William Stansfield (Macmillan, NYC, 1977), p. 85.

p. 152
Philippine Institute of Volcanology and Seismology (Phivolcs), http://phys.org/news/2013-10-philippine-earthquake-miles-long-rocky-wall.html

p. 154
http://www.geology.ohio-state.edu/~vonfrese/gs100/lect04/

p. 162
http://eol.org/data_objects/22848170

p. 163
http://www.amazon.com/Forbidden-Archeology-Hidden-History-Human/dp/0892132949
"300-million-year-old UFO tooth-wheel found in Russian city of Vladivostok" by Yulia Zamanskaya, http://voiceofrussia.com/2013_01_19/300-million-year-old-UFO-tooth-wheel-found-in-Russian-city-of-Vladivostok/
Accessed 10-16-14.

p. 165
Proxylastodoris
http://www.zoologischemededelingen.nl/z/zoomed/images/vol84/nr01/8401a06fig1.jpg

p. 166
http://www.amazon.com/Vertebrate-Fossils-Evolution-Scientific-Concepts/dp/2881249965

p. 167
http://fc04.deviantart.net/fs71/f/2012/175/3/1/mosasaur_by_sangel99-d54on17.jpg
http://www.nbcnews.com/id/19606626/ns/world_news-asia_pacific/t/chinese-villagers-ate-dinosaur-dragon-bones/#.Url69n90bAI

p. 168
http://www.smithsonianmag.com/science-nature/stegosaurus-rhinoceros-or-hoax-40387948/?no-ist
http://www.qsl.net/w5www/mokele.html

p. 169
http://www.imdb.com/title/tt0088760/

p. 170
Ancient Chinese Inventions by Deng Yinke (Cambridge University Press, 2010), p. 35.

p. 171
Mystery in Acambaro: Did Dinosaurs Survive Until Recently? by Charles H. Hapgood (Adventures Unlimited Press, 1999), cover.
http://books.google.com/books?id=gx5aVytf69MC

p. 174
http://www.youtube.com/watch?v=PRZL7tcLgX8

p. 175
Anthraconauta, *The New Catastrophism* by Derek Ager (Cambridge University Press, 1993), p. 135.
http://en.wikipedia.org/wiki/Lepidosiren

p. 181
The War of the Worlds by H.G. Wells (Sourcebooks, Naperville, IL, 2005), p. 9.

p. 182
http://teeic.anl.gov/er/coal/restech/dist/
p. 186
http://science1.nasa.gov/science-news/science-at-nasa/2003/31oct_hermes/

p. 187
http://en.wikipedia.org/wiki/Bill_Gates

Index